Criteria (Dose/Effect Relationships) for

CADMIUM

Published for the Commission of the European Communities,
Directorate General Scientific and Technical Information
and Information Management, Luxembourg

Criteria (Dose/Effect Relationships) for

CADMIUM

Report of a Working Group of Experts
prepared for the Commission of the European Communities,
Directorate-General for Social Affairs,
Health and Safety Directorate

Published for the
COMMISSION OF THE EUROPEAN COMMUNITIES
by
PERGAMON PRESS
OXFORD · NEW YORK · TORONTO · SYDNEY · PARIS · FRANKFURT

U.K.   Pergamon Press Ltd., Headington Hill Hall,
                Oxford OX3 0BW, England

U.S.A.   Pergamon Press Inc., Maxwell House, Fairview Park,
                Elmsford, New York 10523, U.S.A.

CANADA   Pergamon of Canada Ltd., 75 The East Mall,
                Toronto, Ontario, Canada

AUSTRALIA   Pergamon Press (Aust.) Pty. Ltd., 19a Boundary Street,
                Rushcutters Bay, N.S.W. 2011, Australia

FRANCE   Pergamon Press SARL, 24 rue des Ecoles,
                75240 Paris, Cedex 05, France

WEST GERMANY   Pergamon Press GmbH, 6242 Kronberg/Taunus,
                Pferdstrasse 1, Federal Republic of Germany

Copyright © 1978 ECSC, EEC, EAEC, Luxembourg

First edition 1978

Library of Congress Cataloging in Publication Data

Evaluation of the impact of cadmium on the health
of man.

1. Cadmium - Environmental aspects 2. Cadmium -
Toxicology
I. Lauwerys, R II. Commission of the European
Communities
614.7   RA1231.C3   77-30193

EUR 5697

In order to make this volume available as economically and rapidly as possible the authors' typescripts have been reproduced in their original form. This method unfortunately has its typographical limitations but it is hoped that they in no way distract the reader.

Printed in Great Britain by A. Wheaton & Co. Ltd., Exeter

ISBN 0-08-022024-X

CONTENTS

PREFACE

The Programme of Action of the European Communities on the Environment requires that an objective evaluation of the risks to human health and the environment from pollution is carried out.   This necessitates the compilation of as complete a bibliography as possible on the effects of the pollutants under consideration, and a critical analysis of this information, so that, for certain of these pollutants, criteria (exposure/effect relationships) can be determined.

In this Programme of Action a list of pollutants was drawn up for priority investigation.   These pollutants were chosen on the grounds of their toxicity and of the current state of knowledge of their significance in the health and ecological fields.   Cadmium is included in this list.

This work has been undertaken by the Health and Safety Directorate of the Directorate General for Employment and Social Affairs.   Meetings of the group of experts directed by the Health and Safety Directorate have discussed this report and agreed its contents.   This report, for which the rapporteur was Professor R. Lauwerys, is therefore the reference document on which a report to the Council of Ministers will be made.

P. Recht
Director of Health and Safety
Commission of the European Communities

Chapter I

SUMMARY

Cadmium is a non-essential element present as a contaminant in food, water as well as in polluted air. Its acute toxicity by inhalation or ingestion was recognized many years ago.

Its chronic toxicity to workers exposed to dust and vapour is known approximately since 1950. Concern about its possible effect on the general population exposed to low concentrations over long periods of time has been raised in part because of its steadily increasing consumption and release in the environment mainly during the last 20 years, and in part because of the outbreak of Itai-Itai disease in Japan for which cadmium has been shown to be one causative factor.

I - 1   CHEMICAL AND PHYSICAL PROPERTIES OF CADMIUM AND ITS COMPOUNDS

The chemical and physical properties of cadmium are briefly listed in Chapter II.

Two properties have important environmental consequences:
- a. relatively high vapour pressure which explains its loss by vaporization in the environment during thermal treatment;
- b. solubility in weak dilute acids (and soft water).

Unlike alkyl-mercury derivatives, the aklyl-cadmium compounds are unstable and do not constitute an environmental hazard.

I - 2   NATURAL OCCURRENCE, PRODUCTION, USES AND SOURCES OF ENVIRONMENTAL
         POLLUTION

These topics are dealt with in Chapter III.

Cadmium is not found in a pure state in nature but usually follows zinc
deposits.   Cadmium is therefore always present as an impurity in zinc.

Cadmium production began at the end of the nineteenth century.   It is
produced as a by-product of the mining and refining of zinc ores.
Approximately 70% of the total world production has occurred within the
last 20 years.

The principal uses of cadmium are: fabrication of alloys and solders,
plating of metals, pigments and stabilizers in plastic material and
batteries.   It is estimated that less than 10% of cadmium utilized for
these applications is recycled.

Other sources of environmental pollution by cadmium are: primary iron
and steel industry and secondary non ferrous metal industry, discharges
from incinerators, sewage sludge, wear of automobile tyres, phosphate
fertilizers, consumption of fossil fuels, and corrosion of zinc.
Consumption of tobacco may represent an important source of personal
pollution by cadmium.

I - 3   CONCENTRATION OF CADMIUM IN AIR, FOOD AND WATER

Chapter IV summarizes in table form the concentrations of cadmium which have
been found in air, food and water.   The chemical form of cadmium in the environment (in particular in food) is usually not known and analyses have been
restricted to the determination of "total" cadmium content.

a.   Air
     Cadmium levels in rural air are very low ($0.0001-0.043\mu g/m^3$) and even in
     industrialized cities levels are so low ($0.002-0.7\mu g/m^3$) that the average

intake by inhalation is small compared to that from food (see Chapter V).

The air concentration near Cd emission sources may sometimes be much greater ($0.010-5\mu g/m^3$) and the intake by inhalation can then reach a critical level (see Chapter V). Concentration of cadmium in dust fall usually follows the same pattern as Cd concentration in air (rural < residential < industrial areas).

Tobacco contains cadmium and it has been estimated that 0.1 to $0.2\mu g$ Cd can be inhaled per cigarette smoked.

b. Food

In general the concentration of cadmium in most foodstuffs from non-contaminated areas is below 0.1 ppm but liver, kidney and shell fish can contain much higher concentrations.

Some vegetables and cereals concentrate cadmium when cultivated on a polluted soil can contain concentrations around 1 ppm or above.

If we except some marine animals like oysters which have a remarkable ability to concentrate cadmium above the low level in sea water, there is little evidence for a concentration of cadmium in marine food chains. Food can also be contaminated during storage in cadmium ceramic containers.

c. Water

Surface waters that contain more than a few ppb Cd have probably been contaminated by industrial wastes or by the leaching of land fill or soils to which sewage sludge has been added.

Cadmium in drinking water is usually below $5\mu g/litre$. Two potential sources of cadmium in drinking water are the galvanized pipes sometimes used in plumbing and the silver-base solders in drinking water systems where copper pipes are used.

In some areas well water may contain high concentrations of cadmium.

d.   Analytical methods for cadmium measurements

The results of two recent interlaboratory comparison programmes indicate
that effort must be devoted to improving the precision and accuracy of
cadmium determinations in biological material, mainly in blood.

I - 4   METABOLISM

The information available regarding the fate of cadmium in the organism is
reviewed in Chapter V.

4.1 Intake (respiratory and oral intake)

For the general population living in non polluted areas food constitutes
the most important source of cadmium (median 43µg/day, range 4-84µg/day)
Under normal circumstances drinking water and ambient air contribute much
less to the daily intake (water: 3µg/day; air: 0.005-3.5µg/day).

In certain circumstances however (persons living very close to a cadmium
emission source, heavy smokers) and because of a higher absorption rate
from the lung than from the gastrointestinal tract, airborne cadmium may
become an important source of exposure. In Europe and in the USA it is
estimated that the total daily intake by all routes can range from 6-
116µg/day. In certain regions of Japan where the soil is highly contaminated by cadmium and where locally grown rice is consumed the oral
intake can be as high as 600µg/day.

4.2 Absorption

It has been estimated that about 64% of the amount of cadmium deposited
in the lung can be absorbed. In the general environment 20-30% of the
inhaled cadmium is probably deposited in the pulmonary compartment, thus
13-19% of the total amount inhaled is effectively absorbed.

In man the average oral absorption rate of cadmium has been evaluated at
5%. Physiological (age) and nutritional (iron, protein ...) factors ma

modify this absorption rate.   Further research work is however necessary
to evaluate more precisely the oral and pulmonary absorption rates of
cadmium in man.

If we neglect special circumstances, like living very close to a cadmium
emitting source, it is estimated that the total amount of cadmium absorbed daily by an average adult, non occupationally exposed to cadmium,
ranges from 0.36 to 9.8µg, depending on his smoking and dietary habits.

Some results suggest that persons living close to a cadmium emitting
source could absorb as much as 23µg daily but research work is urgently
needed to evaluate the validity of such reports.

4.3   Transport - Distribution and body burden

In blood, cadmium is mainly concentrated in the red blood cells bound to
a low molecular weight protein (metallothionein).   Some data suggest
that cadmium concentrations in blood reflect current exposure but in view
of the analytical difficulties for accurate determination of cadmium in
blood, this conclusion is only tentative.   Cadmium accumulates in the
organism with age (at least until age 50).   Estimates of body burden of
non occupationally exposed adults have ranged from 9.5-40mg.   This value
can be higher in Japan.   The cadmium content of liver and kidney constitutes approximately 50% of the total body burden.   At age 40-60 mean
concentrations of cadmium in the renal cortex in persons from different
countries (excluding exposed workers and persons living in certain areas
of Japan) have been shown to range from 20-60µg/G wet weight.   In tissue, cadmium is bound mainly to metallothionein.

4.4   Excretion

In man, cadmium (which has been absorbed) appears to be mainly excreted
via the urine;   fecal excretion seems to be less important.   In normal
adults the amount excreted daily via the urine is usually inferior to
2µg/day.

It is hypothesized that at low exposure level cadmium in urine reflects
both exposure and body burden while at higher exposure it mostly reflects

current exposure.   Further research work is required to test this hypothesis.   Tubular lesion induced by cadmium is usually associated with a
increased urinary cadmium excretion.

I - 5   TOXICOLOGICAL EFFECTS OF CADMIUM

5.1 Acute and short term exposure

In human beings the two main target organs of cadmium are the
gastro intestinal tract after acute ingestion and the lungs after
inhalation.   The no-effect level of cadmium administered as a
single oral dose to adult man has been estimated at 3mg.   For an
8-hour exposure period the no-effect levels for cadmium oxide fume
and "respirable" dust are probably below 1 and 3 mg/m^3 respectively.

In animals acute administration of cadmium (mainly by the parenteral
route) can produce toxic effects in many organs (kidney, liver,
testes, nervous system, hematopoïetic system, pancreas, cardiovascular
system) but the relevance of these observations for human health is
probably limited.

5.2 Long term exposure

In man the two main target organs after long term exposure to
moderate cadmium concentrations are the lungs and the kidney.   The
available information suggests however that the critical organ
(i.e. the site of the initial lesion) is the kidneys.   Other types
of tissue damage or functional disturbances have also been observed
in man (e.g. bone, hematopoïetic system) and in animals (e.g.
hypertension) and these effects are also reviewed in Chapter VI.

5.2.1   Effect on the lung

Several investigations on cadmium exposed workers suggest that
repeated or prolonged inhalation of cadmium dust or fume produces

an obstructive pulmonary syndrome and emphysema. This toxic
action probably results from the direct action of inhaled cadmium
on the lungs.

The lowest concentration producing this effect is still highly
speculative.

It has been proposed that continuous exposure (24 hours/day for 70 years)
to $2\mu g/m^3$ is probably close to the no-effect level for the lungs. It
should however be recognized that, because of oral intake of cadmium with
the diet, an ambient air exposure which is sufficiently low to prevent
lung damage may contribute significantly to the body burden and hence
favour the development of kidney impairment.

5.2.2   Effect on the kidney

In man cadmium induced kidney lesions have been observed in workers exposed to airborne cadmium and in Japan in persons ingesting contaminated
food (Itai-Itai disease) especially rice grown on soils irrigated with
cadmium polluted water.

In workers kidney lesions usually precede lung damage. The classical
kidney lesion involves the proximal tubule. Since in man the kidney
appears to constitute the critical organ during long term exposure to
cadmium it is of major importance to evaluate the no-effect level of
cadmium on this organ.

Experimental data on animals
The published results of animal experiments are not always comparable because different morphological, physiological and biological procedures
have been selected for evaluating the effect of cadmium on the kidneys.
One publication suggested that in rats 0.2 ppm cadmium in drinking water
for 6 weeks causes a constriction of renal arteries. On the other hand,
another author reported that 24 weeks after the oral administration of 1
ppm cadmium with water no morphological changes could be found by electron microscopic examination of rat kidneys. Contradictory morphological observations have also been reported in rats receiving 10 ppm cadmium
in their diet or drinking water for several weeks. In many long term
experiments, the techniques used for investigating kidney function lacked

sensitivity (e.g. change in protein clearances was not looked for) and
one must therefore conclude that the no-effect level (for the kidney) of
cadmium administered orally to animals is still unknown.   No long term
studies with the use of sensitive methods have been performed to evaluate
the no-effect level of airborne cadmium on the kidney of animals.
Friberg found that 5mg cadmium/m^3 administered to rabbits (3 hours/day)
caused proteinuria after 4 months.

The critical concentration of cadmium in kidney cortex of experimental
animals is also controversial.   Recently it was reported that, in rats,
cadmium concentration in whole kidney below 1 ppm was already associated
with constriction of smaller renal arteries and that tubular lesions
could be found in mice with kidney concentrations below 3 ppm.   Other
workers have found no renal change in rats with a cadmium concentration
in whole kidney of 15 ppm (i.e. ± 20 ppm in renal cortex).

Epidemiological data

The few studies performed among workers exposed to cadmium suggest that
an exposure to a "respirable" cadmium dust concentration of above 20μg/m^3
(8 hours/day for 20 years) could induce kidney lesions after a 20 year
exposure.   If it is extrapolated to a continuous exposure (24 hours/day
for 70 years) this concentration has to be reduced to about 2μg/m^3
(dust <5μ).

Epidemiological surveys performed in Japan suggest that a continuous oral
daily intake of 200μg cadmium could cause an increased prevalence of kidney damage in persons over 50 years (although the amount of cadmium absorbed through the lung by the Japanese living in polluted areas is not
known, it is believed to be negligible in comparison with oral absorption).   It has been estimated that such an intake corresponds to an
average urinary excretion of 6μg cadmium/litre but it should be stressed
that this value is highly tentative and more research work is required
before proposing a biological threshold for cadmium in urine.

If one makes several assumptions regarding deposition, absorption and
distribution of cadmium in the organism following its inhalation one can,
on the basis of the results of two recent epidemiological studies on
workers, estimate that a renal cortex concentration of about 300μg/g wet

weight can be associated with functional signs of kidney lesions. After oral absorption of cadmium, one can also estimate that a kidney cortex concentration of 364-370µg/g wet weight could increase the prevalence of proteinuria, which is not very different from the above estimation based on the examination of workers exposed by inhalation.

The Subcommittee on the Toxicology of Metals under the Permanent Commission and International Association of Occupational Health has recently endorsed Friberg's proposal of 200µg cadmium/g wet weight as the tentative critical concentration in human kidney cortex (344). However, the tentative character of this proposal should be stressed because it is based on few observations on humans and some recent experimental results suggest that kidney impairment could occur at a lower kidney concentration. Considering first the oral intake and neglecting intake via ambient air or smoking, Friberg calculated that the necessary daily cadmium intake to reach the critical concentration of 200 ppm in the kidney cortex at age 50 is 248µg. It can also be evaluated that the adult oral intake of cadmium necessary to reach 50 ppm in renal cortex at age 50 is 62µg/day. Considering only the respiratory intake, the necessary cadmium concentration in ambient air to reach the critical cadmium concentration in kidney cortex at age 50 is $2µg/m^3$ (ventilation $20m^3/24h$.). On the basis of the same model the daily cadmium absorption (by all routes) required to reach 200 ppm in the renal cortex at age 50 is 10µg. However, as Fleischer et al have recently pointed out, although the assumptions made seem valid as working hypotheses, there are many uncertainties in this model:

1. The "tentative value of the critical level" in the target organ has already been stressed. (Some animal experiments suggest that the critical level is much lower than 200 ppm).

2. "The unusual nature of the cadmium binding protein in the kidney makes it uncertain whether overall cadmium levels relate to toxicity".

3. "Analytically valid long-term human balance studies are not available" to precisely estimate the rate of cadmium absorption and excretion.

5.2.3   Effect on the cardiovascular system (in particular hypertensive effect)

Controversial results have been reported regarding the hypertensive action of cadmium in animals and in man.   Some authors have claimed to have induced hypertension in rats (and change in arteriolar vessels of kidneys) by administering cadmium orally at very low doses (0.02 ppm in diet, 0.2 ppm in water).   Other authors could not reproduce these observations.   The existence of a causal relationship between cadmium and hypertension in man is suggested by some epidemiological studies in the general population, but this has not been supported by other studies.

5.2.4   Effect on the bones

Osteomalacia and osteoporosis with a tendency to fracture and bone deformation accompanied by lumbar pains, leg myalgia and pains on bone pressure as well as disturbance of gait have been described in Itai-Itai patients, principally in women after menopause who had borne several children.   The same type of bone lesion (in particular pseudofracture) has also been found in workers exposed to cadmium.   Bone lesions can also be induced in animals treated with cadmium.   Some authors believe that the bone lesion is not the result of a direct action of cadmium but is probably due to a disturbed calcium and phosphorus metabolism secondary to the kidney lesion.   Others are of the opinion that cadmium exerts a direct toxic action on bone tissue which can even precede the kidney damage.   In animals, the lowest cadmium concentration which has been reported to induce changes in bone metabolism is 10 ppm in the diet.

5.2.5   Effect on the hematopoietic system

Slight hypochromic anaemia has been seen among most Itai-Itai patients as well as among workers exposed to cadmium.   Anaemia has also been produced in animals.   In animals the no-effect level of cadmium on the hematopoietic system lies probably between 5 and 10 ppm in drinking water since short term administration of 10 ppm to mice has been found to partially inhibit the gastrointestinal absorption of iron.

5.2.6   Effect on the liver

In workers exposed to cadmium and in Itai-Itai patients, functional liver
changes have rarely been reported.   In animals the lowest cadmium concentration which has induced changes in liver enzyme activities is 1 ppm
in drinking water.   The health significance of these changes is, however, unknown.

5.2.7   Effect on animal growth and survival

The published results concerning the effect of cadmium on animal growth
and survival are contradictory.   The lowest concentration which has been
reported as affecting growth or survival is 3 ppm in drinking water.

5.2.8   Other effects: anosmia, dental caries, pancreatic and adrenal functions, immunosuppression α_1 antitrypsin inhibition

Except for anosmia, the other effects have only been described in animals
or during in vitro experiments (e.g. α_1 antitrypsin inhibition).

5.3 Carcinogenicity

Parenteral administration of cadmium is carcinogenic in rats (local sarcoma and interstitial cell tumour in the testes).   Further epidemiological studies are required to establish whether or not cadmium exhibits a
carcinogenic action in man.

5.4 Mutagenicity

Further work is necessary to evaluate the mutagenic action of cadmium in
man since one study reported an increased frequency of chromosome abnormalities in peripheral leucocytes of Itai-Itai patients while another
similar study could not confirm this finding.   Chromosome analyses of
peripheral lymphocytes of workers exposed to lead and cadmium revealed an
increased number of chromosomal anomalies suggesting a synergistic effect
between both metals.

5.5 Effect on reproductive processes

Doses up to 100 ppm of cadmium in their diet given to male and female rats did not affect the fertility of the animals. One single subcutaneous injection of 0.33 mg $CdCl_2$/kg body weight on day 7 of pregnancy has shown to induce embryolethality in mice; teratogenic effects appeared above 0.63 mg/kg. In rats embryolethality and teratogenicity was seen following one single intraperitoneal injection of 2.5 mg/kg early in pregnancy; however, lower dose levels were not tested. Oral doses of cadmium of >10 mg/kg have been tested only in mice and rats during organogenesis. At these dose levels cadmium also induced embryolethality and/or teratogenicity in these animal species. Insufficient epidemiological data exist to estimate the teratogenic potential of cadmium on humans.

I - 6   STANDARDS

Standards edicted or recommended by some countries or International Agencies are briefly listed in Chapter VII.

I - 7   NEEDS FOR FURTHER RESEARCH

Research recommendations are the object of Chapter VIII. They have been classified in three general areas:
    cadmium analysis, metabolism and toxicity.

Chapter II

CHEMICAL AND PHYSICAL PROPERTIES OF CADMIUM AND ITS COMPOUNDS (119, 134)

Cadmium was discovered in 1817 by Strohmeyer and Herman in Germany. It is a transition metal in group IIb of the periodic table of elements (like zinc and mercury). It is a white metal with a bluish tinge. It is soft, of considerable ductility and easily worked.

Atomic number 48; atomic weight 112.40; specific gravity 8.65; melting point $320.9^{\circ}C$; boiling point $767^{\circ}C$.

There are eight stable isotopes of abundance:
^{106}Cd: 1.22%; ^{108}Cd: 0.88%; ^{110}Cd: 12.39%; ^{111}Cd: 12.75%; ^{112}Cd: 24.07%; ^{113}Cd: 12.26%; ^{114}Cd: 28.86%; ^{116}Cd: 7.58%.

Its vapour pressure is 1.4mm at $400^{\circ}C$ and 16 mm at $500^{\circ}C$ so that losses by vaporization are to be expected during thermal treatment such as ore roasting, brazing, the remelting of steel scrap and the incineration of cadmium-containing refuse. Cadmium shows only valence +2 in its compounds. In air, the vapour oxidizes quickly to produce cadmium oxide. Cadmium dissolves in weak dilute acids, a property which has been responsible for acute oral intoxication in man. The sulphide CdS, the carbonate $CdCO_3$, the oxide (CdO) and the hydroxide ($Cd(OH)_2$) are insoluble in water. Cadmium sulphide is decomposed by acids with the liberation of hydrogen sulphide gas.

The fluoride, chloride, bromide, iodide, nitrate and sulphate of cadmium are relatively soluble compounds. Cadmium forms also a wide variety of soluble complexes, notably with cyanides and amines.

Unlike alkyl-mercury derivatives, the alkyl-cadmium compounds are extremely unstable, reacting rapidly with water and moist air under environmental conditions. Hence they are not expected to be of importance as environmental pollutants.

Chapter III

NATURAL OCCURRENCE, PRODUCTION, USES AND SOURCES OF ENVIRONMENTAL POLLUTION

III - 1   NATURAL OCCURRENCE

Cadmium occurs throughout the lithosphere but is not found in a pure state in nature.   The average concentration of cadmium in the earth's crust is esti-mated at 0.15-0.2 ppm.   It is principally concentrated in sulphide deposits, in which it follows zinc and, to a much lesser extent, lead and copper.   It is therefore found mainly in zinc, lead-zinc and lead-copper-zinc ore.   The amount in the principal zinc ore (ZnS) varies from 0.1-5% and sometimes higher (65).   Commonly the content of cadmium in zinc concentrates ranges from 0.2-0.4%.   The cadmium content of the majority of copper-zinc deposits is 0.3 parts cadmium per 100 parts zinc, and for lead-zinc deposits 0.4 parts cadmium per 100 parts zinc.   Cadmium is therefore always present as an impurity in zinc.   Table 1 summarizes its average concentration in various media (1, 39, 119, 188, 248, 357, 537).

T A B L E   1

Concentration of cadmium in various media

Igneous rocks:	0.001 - 0.6 ppm
Bitumous shales:	<0.3 - 11 ppm
Sandstones:	0.05 ppm
Limestones:	0.035 ppm
Coal:	0.25 - 5 ppm
Oil:	0.01 - 16 ppm
Phosphatic rock:	traces - 100 ppm
Seawater - Atlantic Ocean, Irish Sea, English Channel:	0.05 ppb (0-0.6)
- North Sea:	0.4 ppb (0-1.5)
- British coastal waters:	0.2 - 9 ppb
Fresh water:	<1 - 2 ppb
Marine sediments (Atlantic and Pacific Oceans):	0.1 - 1 ppm
River and lake sediments:	5 - 430 ppm
Soil surface (0-5 cm) average of uncontaminated areas:	0.4 ppm

III - 2   PRODUCTION
         _____

The production of cadmium began towards the end of the nineteenth century.
It is produced commercially only as a by-product of the mining and refining of
zinc ores.   In 1969 in the US, zinc recovery operations accounted for 85% of
the cadmium produced, the remaining 15% of the cadmium was produced from
imported materials (primarily rich zinc dusts) and from scrap metal (511).
Production and consumption of cadmium is continuing to expand throughout the
industrialized world.   The world production of cadmium by decade is summarized in Table 2 (357, 508).

T A B L E   2

Estimated World Production of Cadmium (metric tons)
           (total for ten-year periods)

1910 - 1919 :	977
1920 - 1929 :	5,404
1930 - 1939 :	28,288
1940 - 1949 :	49,094
1950 - 1959 :	74,999
1960 - 1969 :	124,207

Approximately 70% of the total world production of cadmium has occurred within
the last 20 years.   The trend in cadmium production in some European countries is shown in Table 3 (357, 508) and Table 4.

It is evident that primary lead-zinc industry (including mining, concentrating, smelting and refining) and primary cadmium industry (recovery and refining) constitute important sources of environmental pollution by cadmium.
These industries release dust, fumes, waste waters and sludge containing
cadmium.   Modern technology makes it possible to considerably reduce direct
emission from these sources (to $\frac{1}{10}$ - $\frac{1}{100}$ previously recorded levels).

T A B L E   3

Cadmium production (metric tons)
(total for ten year periods)

Country	1910-1919	1920-1929	1930-1939	1940-1949	1950-1959	1960-1969
Austria	-	-	-	-	44	199
Belgium	-	10	2,270	1,108	5,623	7,839
France	-	191	1,216	350	1,401	3,933
Germany	521	86	1,860	1,407	2,149	3,896
Italy	-	-	418	853	1,844	2,636
Netherlands	-	-	-	-	131	697
Norway	-	-	1,204	313	1,014	967
Poland	-	-	1,132	1,803	2,340	4,034
Spain	-	-	-	10	72	556
United Kingdom	-	29	415	1,460	1,377	1,676
Yugoslavia	-	-	-	-	84	718
U.S.S.R.	-	-	115	223	2,290	17,668

T A B L E   4

Cadmium production (metric tons)

Country	1969	1970	1971	1972	1973
Belgium	949	1093	947	1150	1279
France	523	528	579	572	606
Germany	792	1035	982	913	1221
Italy	422	426	350	416	397
Netherlands	111	111	123	122	31
United Kingdom	245	318	262	240	314

(data provided by the Zinc Development Association, London)

III - 3   USES AND SOURCES (OTHER THAN CADMIUM-PRODUCING INDUSTRIES) OF

ENVIRONMENTAL POLLUTION

The amount of cadmium consumed in recent years by some European countries is shown in Table 5.

T A B L E   5

Cadmium consumption (metric tons)

Country	1969	1970	1971	1972	1973
Belgium	676	788	703	1,086	1,357
France	1,190	1,028	969	1,000	1,150
Germany (F.R.)	2,298	1,801	1,788	1,964	2,183
Italy	350	370	320	350	430
United Kingdom	1,473	1,313	1,176	1,354	1,563
World total	18,222	14,542	15,000	17,005	17,813

The principal uses of cadmium are (357):

—   Fabrication of alloys and solders

Cadmium is a common constituent in low-melting alloys, usually in con-
junction with metals such as bismuth, lead and tin.   Cadmium-containing
alloys are used in the production of bearings for aircraft and other
internal combustion engines, for solders and for low-melting and brazing
alloys.   Silver brazing alloys contain silver, cadmium, copper and zinc
and are used quite commonly as contacts in the electrical and electronics
industry.   Silver solders contain cadmium in amounts ranging between 15
and 23%.   A recent use of cadmium is in the production of automobile
radiators containing 0.2% cadmium.   Currently cadmium used in alloys
averages about 4 to 12% of the total consumption (10% in the US in 1972,
11% in the Netherlands in 1955) (71).

-   Plating of metals

Cadmium is coated over iron, steel, copper alloys, aluminium alloys by
electrodeposition, dipping or hot spraying.   Cadmium plating is used for
the following: components for aircraft, components for automobile, electrical and electronic apparatus, household appliances, radio and tele-
vision sets, hardware, fasteners (511).   Depending on the country 10 to
70% of cadmium production is consumed for electroplating (35% and 49% in
the US in 1972 and 1974 respectively, 58% in the Netherlands in 1955, 30%
in the UK in 1974.   The electroplating industry has the largest number
of users of all cadmium consumers.   The platers are widely distributed
with the larger facilities located near industrial areas such as automobile manufacturing centres (511).   The cadmium wastes from electroplating are primarily from the rinsing operations and the typical aqueous
waste stream contains 100 to 500 ppm cadmium in addition to other heavy
metals, cyanides, and metal surface cleaning agents (511).

-   Pigments and stabilizers in plastic material

Compounds of cadmium are used as colourants in various materials as
paints, enamels, ceramic glazes, rubber, glass, textiles, coated fabrics,
leather, printing ink and plastics.   Cadmium sulphide (cadmium yellow)
is probably the most widely used cadmium compound as a pigment.   Complex
cadmium salts containing $CdS-BaSO_4$, $CdS-BaSO_4-ZnS$ and $CdS-BaSO_4-CdSe$ are
referred to as cadmium lithophones and are used to provide a range of
yellow, orange and red pigments.   The cadmium lithophones contain about
15 to 27% cadmium.   Besides the cadmium pigments referred to, the ceramic industry uses, for the colouring of glazings, other cadmium compounds
such as cadmium oxide or cadmium carbonate in the presence of selenium
and sulphides.   Cadmium pigments provide extreme colour retention and
are in demand in exposures where heat resistance is essential.   They are
extensively used in ethylene, styrene and vinyl chloride polymers.
Cadmium salts of long chain fatty acids are also widely used as stabilizers in plastic material principally polyvinyl chloride plastics which
can contain 0.1-0.3% cadmium.   Cadmium used as a pigment and as a plastic stabilizer accounts for about 25-45% of the total consumption (33% in
the US in 1972, 27% in the Netherlands in 1955-1956 (71), 45% in Germany
in 1973 (250), 43.2% in the UK in 1974).

- ## Fabrication of batteries

In the USA approximately 3% of total cadmium consumption was used for
battery production in 1968 but in Germany 12.2% of cadmium processed in
the country in 1973 was for battery production. The use of the nickel-
cadmium battery is expanding but the silver-cadmium cell is in only
limited production. The advantages of the nickel-cadmium battery are
long life, simple maintenance, maximum current density with minimum
voltage drop, quick charging, and the ability to operate over a wide
temperature range. It is utilized in such diverse applications as the
small rechargeable items such as flashlights, electric shavers, and
cordless carving knives, to heavy equipment uses such as buses, diesel
locomotives, aeroplanes and spacecraft. The most rapidly expanding
application for this type of battery is in calculators (511).

- ## Other uses

Small amounts of cadmium are used for the production of fungicides (232),
control rods for nuclear reactors, fluorescent lamps, phosphors for tele-
vision tubes, luminescent dials, compound in photography, lithography,
process engraving, curing rubber and various miscellaneous uses, (e.g.
the fluorescent particle atmospheric tracer which is commonly used in
atmospheric and air pollution studies is a finely powdered mixture of
zinc sulphide and cadmium sulphide) (458).

Other sources of environmental pollution:

- ## Primary iron and steel industry and secondary non-ferrous metals industry

The industry which remelts and recovers plated and galvanized ferrous
metals and non-ferrous metals and alloys produces also dust, fumes, waste
waters and sludge containing cadmium.

- ## Incinerators

The incineration of wastes containing cadmium (plastics, pigments, plated
scrap..) can result in local discharges of cadmium aerosols probably as
CdO. Solid waste stream from incinerators will also contain cadmium.

- #### Wear of automobile tyres

Tyres can contain 20 to 90 ppm cadmium (present as an impurity in the zinc oxide used as a curing accelerator).

- #### Phosphatic fertilizers

Phosphate rocks may contain up to 170 ppm of cadmium.   Thus phosphate fertilizer is also a source of cadmium.   The cadmium content in super-phosphate fertilizer is usually in the range 2-50 ppm (135, 464, 531) but Caro has discovered concentrations as high as 50-170 ppm in superphosphate (308).   Kloke (232) has estimated that the increase in cadmium concentration of soil resulting from the application of phosphatic ferti-lizer (50 kg P_2O_5/ha) could at most be 0.016 ppm per year.   Since the solubility of cadmium phosphate is low, it is usually believed that any cadmium precipitated as phosphate would be relatively unavailable for uptake by plants.   However, if ammoniacal fertilizers are used there is a danger of large quantities of dissolved cadmium resulting from the formation of the soluble cadmium-ammonia complex ions $Cd(NH_3)_2^{+2}$ and $Cd(NH_3)_4^{+2}$ (511).

- #### Consumption of coal and heating oil

We have indicated that cadmium is present as a trace element in fossil fuels.   Cadmium is released in the environment (fume, ash) during con-sumption of these fuels.   The range of cadmium concentration in coal is between 0.25 and 5 ppm.   It is not known, however, what percentage of cadmium in coal is emitted from the stack, retained in the furnace or removed by the scrubbers or precipitators.   Coal ash contains up to 150 ppm cadmium (188).   The average concentration of cadmium in heating oil is approximately 0.3 ppm.

- #### Sewage sludge

Sludge from sewage plants is rather rich in cadmium.   In Sweden the ave-rage cadmium content of the sludge from 56 plants was 15.6 ppm (350). The use of this sludge as fertilizer could increase cadmium concentra-tions in soil (129) and it is known that some plants (rice and wheat for example) can take up considerable quantities of cadmium from soil (276, 463).

Fulkerson (135) has calculated that application of a few tons of sewage sludge per year containing 20 or more ppm cadmium to unpolluted agricultural soils (<0.1 - 0.5 ppm) will raise the concentration of the ploughed layer of the soil to levels approaching 1.2 to 6 ppm.

- Zinc

We have indicated above that cadmium is always present as an impurity in zinc.   At present, redistilled zinc contains 1-10 ppm cadmium but raw zinc has a higher cadmium content (usually about 0.2%).   Zinc with the lowest cadmium content (1-10%) is used mostly for alloys and zinc with the highest cadmium content (0.2%) is used for galvanizing.   It has been estimated that due to corrosion, all of the cadmium in galvanized coatings is released to the environment within 4 to 12 years (511).

- Tobacco

The consumption of tobacco releases cadmium.   This will be discussed in Chapter IV.

Because of its types of application which are highly dissipative little amounts of cadmium (probably less than 10%) are recycled.   It is not, therefore, surprising that it has been estimated that nearly one third of the cadmium used annually finally reaches the environment (327).   According to this last study the most significant sources of emission to the atmosphere are metallurgical processing (smelting operations, remelting of cadmium-plated steel), incineration of metal refuse, coal and oil combustion.   These sources constitute more than 90% of the total cadmium emission to the atmosphere in the USA.   Burning of oil, tyres, PVC-plastics and waste will also contribute. Water pollution would mainly result from mining and beneficiation, hydrometallurgy and electroplating.   The use in agriculture of fertilizers and sewage sludge will contribute to the cadmium load on arable land.   A balance for cadmium in the Federal Republic of Germany (250) in 1973 has been established and is reproduced below.   Direct emission in the environment is considered to represent 5.2% of the total amount of cadmium processed in the country.

MATERIAL BALANCE FOR CADMIUM IN FRG IN 1973 (IN METRIC TONS)
(from ref. 250)

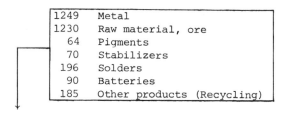

```
                              ┌──────────────────────────────────────────┐
                              │ 1249   Metal                             │
                              │ 1230   Raw material, ore                 │
                              │   64   Pigments                          │
                        ┌─────│   70   Stabilizers                       │
                        │     │  196   Solders                           │
                        │     │   90   Batteries                         │
                        │     │  185   Other products (Recycling)        │
                        │     └──────────────────────────────────────────┘
                        ↓
```

Import 3084

```
                                281   Primary raw material (ore metal)
                                 50   Raw material for solders
                                139   Raw material for metal
                                170   Non specified

                              ┌──────┐
                              │ 3724 │
                              └──────┘
                                 ↓
                    ┌───────────────────────┐
                    │ Cadmium processed     │          189 (5.1%)
                    │ stockpile + 116       │
                    └───────────────────────┘        ┌──────────────┐
                                        386          │  Recycling   │
                                                     └──────────────┘
    2054 (55.2%)                      ┌──────────┐
                                      │ Emission │    ┌──────────────┐
┌─────────────────────────────────┐  197 (5.2%) │    │  66  water   │
│ 719 Pigments (19.3%)            │  └──────────┘    │   7  air     │
│ 349 Stabilizers (9.4%)          │                  │ 124          │
│ 335 Batteries (9.0%)            │                  └──────────────┘
│ 386 Galvanized products (10.4%) │
│  55 Glass industry              │
│  54 Alloys                      │
│  15 Rectifiers                  │
│  39 Other products              │
│  43 Cadmium salts               │
│  59 Impurities in zinc          │
└─────────────────────────────────┘
            ↓
        dispersion
```

1168 - Export (32.4%)

----

245 Metal (6.6%)
479 Pigments (12.9%)
129 Stabilizers (3.5%)
120 Batteries (3.2%)
 25 Rectifiers
 11 Alloys
 39 Glass industries
 51 Intermediate products
 56 Cadmium salts
 13 Impurities in zinc

Chapter IV

CONCENTRATION OF CADMIUM IN THE ENVIRONMENT IN PARTICULAR IN AIR, FOOD, WATER

It should first be recognized that the chemical form of cadmium in the environment (in particular food) is usually not known and analyses have therefore been restricted to determination of "total" cadmium content.

IV - 1   AIR

Measurements of the cadmium concentration in air have been made in many countries.   Table 6 summarizes the results found in rural, urban and industrial areas.

T A B L E   6

Cadmium concentration in air

1.   Rural areas

Location	Concentration ($\mu g/m^3$) usual annual average	References
Sweden	0.0009 (monthly)	129
F.R. Germany (Erlangen)	0.0015	103
Tuxedo N.Y.	0.003	233
30 non urban areas in US (1966)		
in 21	0.004	325
in 9	0.004-0.012	-
Northern Norway	0.0001	174
Jungfraujoch (Switzerland)	0.00028	174
Belgium (Botrange)	0.006 (0-0.043)	326
Belgium (2 rural areas) (one day average)	0.00096-0.0059	174, 188

T A B L E 6 (continued)

2. Urban areas

Location	Concentration ($\mu g/m^3$) usually annual average	References
Stockholm	0.005 (weekly)	129
Polish towns	0.002 - 0.051	206
Belgium: 5 residential areas		
(one day average)	0.0029 - 0.507	174
Belgium: Brussels North	0.018 (0 - 0.351)	326
Brussels Centre	0.017	
Brussels South	0.017 (0 - 0.235)	
Ghent	0.0012 - 0.022	86
Tokyo (1969-1970)	0.010 - 0.053	129
	(max. 24 h/0.53)	
Lower Manhattan NY (1970)	0.023	233
Bronx NY (1970)	0.014	233
NASN Station NY (1969)	0.014	129
Chicago and NW Indiana (1971)	0.005 - 0.08 (mean 0.019)	168
Cincinnati (1968)	0.02 - 0.08	257
Cincinnati (1972)	0.0017 - 0.0034	76
136 US Cities (1966)		325
in 96	<0.01	
in 40	0.01 - 0.09	
20 largest US Cities (1969)	0.006 - 0.036	129
Munich, F.R. Germany	0.0069 - 0.011	419
Paris, France	0.006 - 0.060	116

3. Industrial areas

Sweden: 500m from source	0.30 (weekly)	129
100m from source	0.60 (monthly)	129
100m from source	5.40 (max. 24 hours)	129

T A B L E 6 (continued)

3. Industrial areas (continued)

Location	Concentration ($\mu g/m^3$)	References
Japan:		
100 m from Zn smelter	0.50 (weekly)	537
400 m from Zn smelter	0.20 (weekly)	537
500 m from Zn smelter	0.56 (3 day mean)	129
500 m from source	0.16 - 0.32 (8 hour)	129
Osaka	0.015 0.036	537
Annaka (nr Zn smelter)	0.055 - 0.166	540
England, 1km from Zn smelter	0.533	537
U.S.: El Paso (1964)	0.12	129
East Helena (Montana)		
800 m from smelter	0.29 (3 months average)	
	(maximum 0.69)	129
1300m from smelter	0.06 (3 months average)	129
Belgium: Ghent (1971-1973)	<0.010 - 0.2	174
Antwerp (1972 - 1973)		326
North	0.015 (0 - 0.153)	
Centre	0.021 (0 - 0.146)	
South	0.017 (0 - 0.116)	
Charleroi (1972-1973)		
North	0.016 (0.001 - 0.098)	
Centre	0.015 (0 - 0.092)	
South	0.010 (0 - 0.054)	
Liège (1972 - 1973)		
North	0.045 (0.001 - 0.552)	
Centre	0.075 (0.002 - 0.715)	
South	0.034 (0 - 0.354)	
Germany: Stolberg 1973	0.024 - 0.102	313
1974	0.020 - 0.061	
Ruhr area		
10 cities 1974	0.009 - 0.056	314
Around Zn smelter (1971-1975)	0.05 - 0.02	560

It should be stressed that these results are not directly comparable for several reasons (different sampling times, different positions of the sampling
head, different analytical methods).   It is not known precisely in which form
cadmium exists in air.   Suspended aerosol particles are probably composed of
oxides, chlorides and sulphates (183).   Particle size distribution has also
rarely been determined.

At the same time as they measured Cd concentration in air in Cincinatti, Lee
et al (257, 258) determined the mass median diameters of the particles which
averaged 3.1 and 10μ downtown and in a suburb respectively.   In both areas
about 40% of the particles were below 2μ.   In St. Louis where the cadmium
concentration in 1970 was $0.01 \mu g/m^3$, the mass median diameter was 1.54μ.   In
Liège (Belgium) Heindryckx (174) found that Cd was mostly associated with particles of diameter 1-2μ.   Dorn et al (555) found the following size distribution of cadmium particles in two sites - one at approximately 800 m from the
base of a lead smelter stack and the other in a rural area:

Particle diameter	Smelter area (%)	Rural area (%)
> 11	3.63	10.81
7 - 11	2.82	13.51
4.7 - 7	5.24	10.81
3.3 - 4.7	5.65	10.81
2.1 - 3.3	4.44	10.81
1.1 - 2.1	25.81	13.51
0.65- 1.1	28.63	16.22
0.43- 0.65	23.77	13.51

About 40 to 70% of the particles were below 2μ, thus confirming the data of
Lee et al.(257, 258).   Particle sizes are usually smaller in the general
environment than inside factories and this should be taken into consideration
when comparing concentrations inside and outside plants and when estimating
the amount of dust deposited in the lung.

The information presented in Table 6 tends to point out that cadmium concentrations of air in urban areas normally exceed those in non urban areas.   In
areas around cadmium-emitting factories, cadmium concentrations in air several
hundred times greater than those in non-contaminated areas are usually found.
In these areas, soils show also much higher concentrations of cadmium than the
normal content of about 0.4 ppm (119).

To evaluate whether there is a trend in airborne cadmium concentration, all the cadmium data from both urban and non-urban NASN (325) stations collected in the US from 1966 to 1969 have been summarized (Table 7). The table illustrates that even though the number of sites has increased over the time period, the cumulative percentage within an interval by year has remained essentially the same.

In summary cadmium levels in rural air are very low $(0.0001-0.043\mu g/m^3)$ and even in industrialized cities levels are so low $(0.002-0.7\mu g/m^3)$ that the average intake by inhalation is small compared to that from food (see Chapter V). The air concentration near Cd emission sources may be much greater $(0.010-5\mu g/m^3)$ and the intake by inhalation can then reach a critical level (see Chapter V).

T A B L E   7

Number of stations within selected cadmium concentration
intervals 1966 through 1969

Number and percent of sites		Concentration interval $\mu g/m^3$				
		0.010	0.011-0.020	0.021-0.030	0.030	total
1966	Number	116	7	3	1	127
	percent	91	5.5	2.5	1	
1967	number	119	7	2	5	133
	percent	90	6	1.5	2.5	
1968	number	175	2	2	0	179
	percent	98	1	1	0	
1969	number	169	31	5	3	208
	percent	81	15	2.5	1.5	

(from EPA 1974) (325)

ANNEX A   Cigarettes

Tobacco and paper contain cadmium which enters air upon combustion of cigarettes.   Analysing eight brands of cigarettes, Szadkowski et al (478) found a
mean content of 1.4µg per cigarette.   Nandi et al (324) tested six brands of
cigarettes and found a range of 18 to 28.5 and a mean of 22.7µg cadmium per 20
cigarettes.   Mendel et al (305) reported values between 1.56 and 1.96 µg cad
mium per cigarette.   Szadkowski et al (478) have estimated that about 10% of
the cadmium content of the cigarette is inhaled.   Thus 0.1-0.2µg Cd can be
inhaled per cigarette smoked (129).   Westcott and Spencer (526) analysed
cigarettes from 7 countries and found that the cadmium levels were in the
ranges 0.5 to 3.5µg/g tobacco.   The samples of Canadian and Japanese tobaccos
contained more cadmium than the other samples.   The levels of cadmium in
particulate phase smoke from single grade tobaccos and European commercial
cigarettes were generally less than 0.25µg per cigarette (plain cigarette:
0.185µg/cigarette in particulate matter + 0.05 in vapour phase;   filter cigarette 0.06 to 0.10 in particulate phase + 0.007 in vapour phase) and these
values support results reported by Szadkowski et al (478).

Dust

Laamanen (247) measured the concentration of cadmium in dustfall in a rural subarctic area of Finland, and found that the deposition varied from less than 0.001 mg/m^2 month to 0.006 mg/m^2 month. Hunt et al (190) determined monthly concentrations of cadmium in dustfall from residential, commercial, and industrial areas in 77 midwestern cities in the United States over a four-month period (September through December). The concentration in dusts from the residential area (0.040 mg/m^2 month) was significantly lower than that from commercial (0.063 mg/m^2 month) and industrial (0.075 mg/m^2 month) areas. The data reported by Hunt et al (190) permits computations of the concentrations in dusts which were about 13, 12 and 11μg/g for the residential, commercial and industrial areas, respectively. These values are 10 to 100 times greater than the cadmium concentrations of surface soils for the region studied.

In the Nordenham area in Germany the annual average dustfall in 1972 ranged from 1.122 mg/m^2 month in the industrial area, to 0.231 and 0.111 mg/m^2 month at 2 and 12 km respectively from the centre of the industrial area (329). Monthly deposition of cadmium 100m south of a cadmium emitting factory in Sweden for periods from May through October 1970 ranged from 7.7 to 40 mg/m^2 month (Olofsson 1970 cited in ref. 129). The mean monthly value, excluding a period when the plant was completely or nearly closed, 100m from the plant, was 18.1 mg/m^2 month. In Japan, mean values 500 and 1,400 m from a factory over a six-month period were observed to be 6.2 and 1.8 mg/m^2 month, respectively (129).

IV - 2   FOOD

The determination of the cadmium content of several food items has been performed by several authors in different countries.   Examples of cadmium concentrations found in some basic foodstuffs (excluding those obtained in the
polluted areas of Japan) are summarized in Table 8.   This data is given to
appreciate the range of cadmium concentrations usually found in various food
items.   They do not always have the same significance - some results are
based on a few determinations, while others have been obtained from samples
collected in particularly polluted areas and are not representative of the
general situation in the corresponding country.   Some results have been
reported without indicating the method of sample collection.   Furthermore,
differences in the accuracy of the analytical methods may probably also explain some discrepancies.   It is felt, however, that the results are sufficiently numerous to estimate with some degree of confidence the range of
cadmium concentrations found in most foodstuffs.

T A B L E   8

Examples of cadmium content of several foodstuffs

Food	Cadmium content (ppm wet weight)	Country	References
Cereals & Vegetables			
Mean various cereals	0.03	U.K.	499
Wheat flour	0.07	U.S.A.	422
Wheat flour	0.05 - 0.06	New Zealand	159
Wheat flour	0.047	F.R. Germany	103
Wheat flour	0.02	Czechoslovakia	263
Wheat flour	0.025	Japan	194
Wheat flour	0.05 - 0.10	Canada	550
Wheat flour	0.029 - 0.108	Sweden	228
Wheat flour	0.15*	U.S.A.	432

*For note see page 38.

TABLE 8 (continued)

Food	Cadmium content (ppm wet weight)	Country	References
Cereals & Vegetables (cont.)			
Wheat grain	0.01 - 0.04	Australia	312
Oats	0.005 - 0.046	Sweden	228
Corn	0.12	U.S.A.	228
Corn	0	Roumania	406
Barley	0.008 - 0.039	Sweden	228
Barley flour	0.04	New Zealand	159
grain	0.04	New Zealand	159
Rice unpolished	0.04	New Zealand	159
polished	0.02	New Zealand	159
Bread, white	0.22*	U.S.A.	432
Bread, white	0.042	Roumania	406
Bread, white	0.046	Japan	194
Bread, brown	0.05	New Zealand	159
Bread, white	0.03	New Zealand	159
Bread, dark	0.15*	U.S.A.	432
Bread, dark	0.054	Roumania	406
Mean various vegetables	0.04 (0.01 - 0.22)	U.K.	496
Mean various vegetables	<0.02	U.K.	499
Potatoes	0.03*	U.S.A.	432
Potatoes	0.02 - 0.05	New Zealand	159
Potatoes	0.017	Roumania	406
Potatoes	0.038	Japan	194
Potatoes	0.039	F.R. Germany	103
Potatoes	0.09	Czechoslovakia	263
Potatoes	0.08	U.K.	495
Carrots	0.30*	U.S.A.	432
Carrots	0.016 - 0.088	F.R. Germany	375
Carrots	0	Roumania	406
Carrots	0.05	New Zealand	159
Carrots	0.041	Japan	194
Tomatoes	0.03*	U.S.A.	432
Tomatoes	0.01 - 0.02	New Zealand	159
Tomatoes	0.013	Roumania	406

*For note see page 38.

T A B L E   8 (continued)

Food	Cadmium content	Country	References
Cereals & Vegetables (cont.)			
Tomatoes	0.032	Japan	194
Tomatoes	0.015	F.R. Germany	103
Tomatoes	0.06	U.K.	495
Lentils, brown dried	0.06	New Zealand	159
Beans	0.019 - 0.075	F.R. Germany	375
Beans, haricot	0.06	New Zealand	159
Cabbage	0.022 - 0.094	F.R. Germany	375
Cabbage, green	0.01	New Zealand	159
Brussels sprouts	0.05	New Zealand	159
Parsley	0.043 - 0.170	F.R. Germany	375
Radishes	0.011 - 0.027	F.R. Germany	375
Radishes	0.02	New Zealand	159
Rhubarb	0.010 - 0.057	F.R. Germany	375
Lettuce	0.031 - 0.198	F.R. Germany	375
Lettuce	0.02	New Zealand	159
Spinach	0.055 - 0.063	F.R. Germany	375
Spinach	0.02	New Zealand	159
Onion	0.03	New Zealand	159
Onion	0.018 - 0.040	New Zealand	159
Processed green peas	0.01	Australia	312
Peas	0.05	New Zealand	159
Asparagus	0.01	New Zealand	159
Cauliflower	0.02	New Zealand	159
Celery	0.03	New Zealand	159
Cucumber	<0.01	New Zealand	159

Fruit

Mean of various fruit	<0.01	U.K.	499
Unspecified veg. & fruit	0.16	Belgium	39
Banana	0.03	U.S.A.	432
Banana	0.011	Japan	194
Apples	0.0065	Roumania	406
Apples	0.004	Japan	194

T A B L E   8 (continued)

Food	Cadmium content	Country	References
Fruit (cont.)			
Apples	0.005 - 0.027	F.R. Germany	375
Apples	0.01 - 0.02	New Zealand	159
Pears	0.010 - 0.013	F.R. Germany	375
Pears	<0.01 - 0.02	New Zealand	159
Strawberries	0.033 - 0.034	F.R. Germany	375
Red currant	0.036 - 0.101	F.R. Germany	375
Black currant	0.055 - 0.098	F.R. Germany	375
Prunes	0.014 - 0.067	F.R. Germany	375
Prunes	<0.01 - 0.01	New Zealand	159
Sour cherry	0.022 - 0.076	F.R. Germany	375
Gooseberry	0.023 - 0.051	F.R. Germany	375
Oranges	0.01	New Zealand	159
Peaches	<0.01 - 0.01	New Zealand	159
Dairy Products			
Milk	0.12*	U.S.A.	432
Milk	0.004	U.S.A.	91
Milk	0	Roumania	406
Milk	0.009	F.R. Germany	103
Milk	0.010 - 0.076	F.R. Germany	375
Milk	0.01	Czechoslovakia	263
Milk	0.003	Japan	194
Dairy products	0.04 - 0.30	U.S.A.	499
Dairy products	0.002	U.K.	499
Butter	0.02	New Zealand	159
Eggs, whole	0.04	New Zealand	159
Eggs, yolk	0.02	New Zealand	159
Meats			
All types of meat	<0.02	U.K.	499
Not specified	0 - 0.2	F.R. Germany	39
Not specified	0.01 - 0.05	U.S.A.	499

*For note see page 38.

T A B L E   8 (continued)

Food	Cadmium content	Country	References
Meats (cont.)			
Beef	0.89*	U.S.A.	432
Beef	0.02 - 0.10	New Zealand	159
Beef	0.060	Roumania	406
Beef	0.054	Japan	194
Beef	0.005	F.R. Germany	419
Lamb and Mutton chops	0.01 - 0.03	New Zealand	159
Pork, bacon	0.10	New Zealand	159
chops	0.04	New Zealand	159
ham	0.06	New Zealand	159
loin, roasting	0.03	New Zealand	159
sausage	0.07	New Zealand	159
Chicken	0.03 - 0.08	New Zealand	159
Chicken	1.25	U.S.A.	432
Chicken	0.027	Japan	194
Kidney (beef)	0.4 - 1.2	Australia	312
Kidney	0.52*	U.S.A.	432
Kidney	0.43	Roumania	406
Kidney (beef)	0.17 - 0.27	New Zealand	159
Kidney (mutton)	0.04 - 0.13	New Zealand	159
Kidney (pork)	0.07 - 0.18	New Zealand	159
Kidney (beef)	0.27	F.R. Germany	313
Kidney (beef)	4.10	F.R.G. (Stolberg)	313
Kidney (beef)	1 - 1.11	F.R.G. (Nordenham)	329
Kidney (control horse)	35.0	U.K.	149
Kidney (horse from ind. area)	330	U.K.	149
Kidney	12 (up to 40)	F.R. Germany	245
Kidney	0.378 (up to 1.41)	F.R. Germany	419
Kidney (elk)	8.0	Finland (Poorvo)	200
Liver (elk)	1.5	Finland (Poorvo)	200
Liver (control horse)	1.6	U.K.	149
Liver (horse from ind. area)	7.5	U.K.	149
Liver	0.08 (up to 0.3)	F.R. Germany	419
Liver (beef)	0.09	New Zealand	159
Liver (mutton)	0.09 - 0.16	New Zealand	159

*For note see page 38.

T A B L E   8 (continued)

Food	Cadmium content	Country	References
Seafood			
Mean for various fish	<0.02	U.K.	499
13 different species	<0.01 - 0.06	Belgium	399
Various fish in Mediterranean	0.11 - 4.82	France	12
Muscles of various fish	0.08 - 1.67	U.K.	366
Oysters	3.60*	U.S.A.	432
Oysters	0.1 - 7.8	U.S.A. (East)	396
Oysters	0.2 - 2.1	U.S.A. (West)	396
Oysters	0.62	Japan	194
Oysters	0.2 - 34.5	Tasmania	405
Oysters	<0.05 - 0.85	Australia	312
Oysters (canned)	3.31	New Zealand	159
Clams	0.31 *	U.S.A.	432
Clams	0.28	Japan	194
Crab	5 - 33.1	U.K.	366
Various seafish (cod, dogfish, grey mullet, rocking, shote, whiting, ray)	0.1 - 0.6	Europe	39
Crab	0.5 - 10	U.K.	408
Molluscs	2 - 50	Europe	39
Crab	22	Europe	39
Cod	0.001 - 0.041	Norway	171
Cod	0.03 - 0.07	New Zealand	159
Pollack	0.002 - 0.008	Norway	171
Flounder	0.005 - 0.024	Norway	171
Tuna	<0.2	U.K.	484
Eel	0.1 - 0.2	Belgium	249
Eel	0.04	New Zealand	159
Pike, flesh	0.002 - 0.003 ⎫	Finland	
liver	0.003 - 0.055 ⎬	(Gulf of	200
kidney	0.028 - 0.232 ⎭	Finland)	
Pike, flesh	0.004 - 0.005 ⎫	Finland (Gulf of	
liver	0.034 - 0.113 ⎬	Bothnia), slightly	200
kidney	0.169 - 0.339 ⎭	polluted area	
Sole, fillets	0.06	New Zealand	159

*For note see page 38.

T A B L E   8 (continued)

Food	Cadmium content	Country	References
Freshwater Fish	0.003	Japan	194
	0.094	U.S.A.	282
	0.02	U.S.A.	281
	0.2 - 1.2	Europe	39
Sugar	0.4 - 2.0	New Zealand	159
Spices	10% <0.004	F.R. Germany	37
	60% <0.1	F.R. Germany	37
	30% 0.1 - 0.5	F.R. Germany	37

*   The method of Schroeder was found to give results that were too high by a factor of about 2.
The cadmium content of wines has been investigated in Germany.   Values between 0.001 and 0.071 ppm were found (29, 102, 103).

In general it is apparent that the concentration of cadmium in most foodstuffs from non contaminated areas is below 0.1ppm but liver, kidney and shellfish can contain much higher concentrations.   Some vegetables and cereals concentrate cadmium when cultivated on a polluted soil.   In some areas of Japan known to be contaminated by cadmium concentrations around 1ppm or above (up to 4.17ppm) have been found in rice and wheat (540).   Kjellström et al (229) have analysed 75 wheat samples (spring wheat and autumn wheat) harvested from the same "non polluted" farm fields in Sweden between the years 1916 and 1972. The individual scatter was large but a significant (p <0.05) tendency toward increasing concentration with time was evident for autumn wheat.   A similar tendency was found for spring wheat, though not a statistical significance. They suggest that this increase could be explained by general air contamination or the cumulative effect of fertilizers.   If we except some marine animals like oysters which have a remarkable ability to concentrate cadmium above the low level in sea water, there is little evidence for a concentration of cadmium in marine food chains.   The studies which have attempted to evaluate the migration into food of cadmium contained in ceramic household containers have been recently reviewed by Engberg and Bro-Rasmussen (99).   On the basis of the estimates of the amount of Cd which can migrate from ceramic household containers in different foods and the composition of the typical Danish diet they calculated that when the food is stored in utensils fulfilling the Danish test limits (see Chapter VII) the total migration of cadmium into food per person per day could amount to 55µg.

IV - 3   WATER
      ‾‾‾‾

We have indicated above that average cadmium content in pure seawater is approximately 0.05μg/l.   Cadmium concentration in uncontaminated surface water
is usually below 2μg/l (357).

In areas of zinc-bearing formations the concentration is higher (up to 10μg/l)
(1).   The major portion of cadmium is found in suspended particles and in the
bottom sediments.   A study by the U.S. Geological Survey in which 720 water
samples were collected from lakes and rivers throughout the US showed Cd present in 42% of the samples (95).   Its concentration ranged from 1-10μg/l.
In about 4% of the river samples, cadmium was detected in excess of 10μg/l.
The highest concentrations of cadmium in water were usually found in areas of
high population density (134).   In a study sponsored by CIPS in Belgium
(399), the following distribution of cadmium concentration in 480 surface
water samples has been found:

Cd μg/l	% of samples
<1	35.4
1 - 5	22.5
6 - 10	10.8
11 - 20	13.8
21 - 30	7.7
31 - 40	6.2
41 - 50	3.6

According to Fleisher et al (119), the surface waters that contain more than a
few ppb Cd near urban areas have almost certainly been contaminated by industrial waste or by the leaching of land fill or soils to which sewage sludge
has been added.   For example, in waste waters from an electrolytic plating
plant Doolan and Smythe (90) found a concentration of 28.5 ppm.   In samples
from a site 500 metres downstream from a cadmium-emitting factory, the cadmium
content of the water was found to be 4 ppm while the bottom sediments contained 80 ppm (dry weight) (100).

Bouquiaux (39) has summarized the results obtained in the Member States of the European Community. In pure river water the Cd concentration ranged from 0 - 1.2µg/l. In slightly contaminated river water (e.g. former mining area) the concentrations ranged from 0 - 10µg/l (more frequently 2 - 3µg/l). In special cases (river polluted by industrial effluents) the concentrations exceeded 10 µg/l (e.g. in the river Rhine up to 16.2µg/l (222), in some Belgian rivers up to 50µg/l (399)).

Cadmium in drinking water is usually below 5µg/l (39, 499).

A survey conducted in 1969 by the US Public Health Service during which 2595 community water samples were analysed found that the US drinking water standard for cadmium (10µg/l) (400) was exceeded on 0.2% of the water samples. The highest value found was 110µg Cd/litre (298). In some areas well water may contain high concentrations of Cd e.g. up to 225µg/l in one Cd polluted area in Japan (239) and up to 71µg/l in California (449). According to Zoeteman and Brinkmann (551), the average concentration of cadmium in water delivered by public water supplies in 7 large cities of the European Communities amounts to 1.1µg/l with a range of 0.2 - 4.0µg/l. It should be realized that these values apply mainly to the water quality at the pumping station. Therefore the actual present concentrations of these metals at the tap can be higher than suspected.

A potential source of cadmium in drinking water is the galvanized pipes used sometimes in plumbing (432). The fate of the cadmium corroded from the surface of galvanized piping should depend on the carbonate content and pH of the water. If the water is soft and somewhat acid the cadmium could remain in solution.

Another source is the use of silverbase solders in drinking water systems (98) where copper pipes are used instead of galvanized iron pipes. This has been recently confirmed in France where it has been found that samples taken directly from the water reservoirs contained always less than 5µg Cd/l but at the taps 15% of collected samples contained more than 5µg Cd/l (126). In a city in Norway, Stegavid (462) found that in the majority of water samples taken at the taps, cadmium could not be detected.

In 8 samples of bottled mineral waters in the European Communities an average cadmium concentration of 0.32µg/l was found with a range between 0.05 and 1.0 µg/l (551).

ANNEX   Analytical methods for cadmium

                                   measurement

As mentioned above the techniques available for measuring cadmium in the environment and biological materials are not able to differentiate the types of
compounds.

At the present time, the concentration or amount of cadmium in water, air,
soil, plants and other environmental or biological materials is determined as
the element.   Determination of Cd in air, water, food and organs has been
performed by different methods mainly colorimetry (dithizone method), emission
spectroscopy, atomic absorption, spectrophotometry (flame and flameless), ano-
dic stripping, voltametry, atomic fluorescence spectrometry, X-ray fluorescence method, neutron activation analysis, polarographic and potentiometric
methods and fluorimetry.   Various pretreatments of the samples (mineraliza-
tion, chelation, extraction, ion exchange chromatography ...) are usually
required.   Currently atomic absorption spectrometry is certainly the most
widely used method.   We do not intend to describe all these methods.   We
want only to stress the point that precise determination of Cd is as difficult
as is lead measurement.   The few interlaboratory comparison programmes which
have been carried out suggest that many laboratories have not yet mastered the
analytical difficulties for accurate determination in biological materials
mainly in blood (255, 277).

A neutron activation method has been developed for Cd determination in organs
in vivo, mainly in liver (169, 303).   The original method which requires a
cyclotron as a source of neutrons does not lend itself to large population
screening but a small portable neutron activation analysis system has now been
developed (496) which seems very promising.

Chapter V

METABOLISM (INTAKE —
ABSORPTION — DISTRIBUTION —
EXCRETION — BODY BURDEN AND
BIOLOGICAL INDICES OF EXPOSURE
OR BODY BURDEN)

V - 1   INTAKE

The lungs and the gastrointestinal tract constitute the two main routes of intake and absorption of cadmium, the skin plays a negligible role.

1.1 By Inhalation

The amount of cadmium deposited in the lung compartment (Q inhaled - Q
exhaled) depends mainly on the air concentration and on the particle size
(we voluntarily neglect here differences in anatomical and physiological
parameters to consider a "standard lung model" as proposed by the Task
Group on Lung Dynamics 1966: minute volume = 20 l/minute) (482).

The deposition rate in the pulmonary compartment (alveolar bed) varies
inversely with particle size with values of approximately 50% for particles having a mean mass diameter of 0.1μ and approximately 20% for
particles having a mean mass diameter of 2 microns.   The application of
the "standard lung mode" to the cadmium particle size distribution reported in Chapter IV indicates that approximately 20-30% of the cadmium
inhaled in the ambient air would be deposited in the pulmonary compartment.   If we apply this average rate of deposition (25%) to the lowest
and highest ambient concentrations of Cd found in rural, urban and
industrial areas and if we assume an average daily inhalation of $20m^3$ the
amount of Cd deposited in the lower respiratory tract can be estimated at
    rural areas: 0.0005 to 0.215μg/day
    urban areas: 0.01   to 3.5 μg/day
    industrialized areas (cadmium emitting factors): 0.05 to 25μg/day.

It must be recognized that the highest level (25μg) is probably found
only in very exceptional circumstances (proximity of a cadmium smelter).

43

In smokers one must add to this amount the quantity taken up with cigarette smoke i.e. 2-4µg per 20 cigarettes from which approximately 50% can
deposit in the lung (particle size <0.3µ).

1.2 By the Gastrointestinal Tract

Three methods have been used to estimate the intake of Cd with food and
water.

1. Establish a standard diet for the population by determining the na-
    ture and the amount of foodstuffs and beverages ingested. By mea-
    suring their cadmium concentration the total daily intake of the
    metal can be estimated.

2. Collect duplicate meals and analyse them for cadmium.

3. Determine the daily fecal excretion of cadmium. Since it is as-
    sumed that about 5% of ingested cadmium is absorbed (see next sec-
    tion) the daily intake can then be estimated. This method assumes
    also that there is no important gastrointestinal excretion of the
    metal. The uncertainty of both hypotheses must be stressed.

The results obtained by several investigators using one of these methods
are summarized in Tables 9 a and b.

It should be stressed that these estimates apply to an "average adult".
As Tolan (499) has pointed out they are of limited use in assessing exposure of groups of the population most at risk. They do not apply to
areas of Japan where the soil has been found highly contaminated by cadmium and where oral intake (mainly through contaminated rice) has been
found as high as 600µg/day (see Chapter VI). To the results presented
in Table 9a should be added the contribution of water to the amount
ingested with food. In the absence of particular contamination the
cadmium concentration in water is around 2µg/l (or even lower) which
corresponds to a maximum additional intake of <2-4µg/day. When tap
water is contaminated by cadmium, the amount of cadmium ingested with
water can then exceed 10µg/day. If we exclude the Japanese data (since
Japan may be a special risk area) the oral daily intake of cadmium in non
polluted areas ranges between 6 and 94µg with a median of approximately
43µg/day with food and 3µg with water.

T A B L E   9 a

Cadmium intake with food

(standard diet or duplicate meal method)

Country	Cadmium intake μg/day	References
Roumania	38-64	406
F.R. Germany	48	103
F.R. Germany	30	329
F.R. Germany (Nordenham: 1-3 kms from emission source)	43-84	329
France	20-30	116
Japan (non polluted area)	60	540
Japan (non polluted area)	60-120	542
	99-108	542
	47	132
	25-43	541
	31	542
Czechoslovakia	60	263
U.S.A.	4-60	422
	38-50	93
	27-64 (range of average values)	322
	26-50	499
	51.2 (1973)	121
	71.4 (1974)	121
Sweden	10	527
	17.2	559
New Zealand	21	158
Canada	80	499
U.K.	15-30	499

T A B L E 9 b

Cadmium intake with food and water

(fecal excretion method)

Country	Cadmium intake μg/day	References
Japan	$57 \times \frac{100}{95} = 60$	505
F.R. Germany	$31 \times \frac{100}{95} = 32.4$	103
F.R. Germany	$31 \times \frac{100}{95} = 32.4$	480
U.S.A.	$42 \times \frac{100}{95} = 44.2$	498
Sweden (smokers)	$19.1 \times \frac{100}{95} = 20.1$	559
(non smokers)	$15.9 \times \frac{100}{95} = 16.7$	

(The results obtained by Schroeder et al 1976 (432) are omitted because they
are probably erroneously high due to interference in the analytical procedure
selected: direct atomic absorption spectrophotometry without extraction) (129).

1.3 Total Intake

We can estimate that the total daily intake from all sources (excluding
Japan) can range from 6μg for a non smoker living in the least polluted
rural area and eating the less contaminated food (0.0005 from air and 6μg
from food and water) to 116μg for a 20 cigarettes/day smoker living close
to a cadmium emitting source and eating more contaminated food (25μg from
air, 4μg from cigarettes and 87μg from food and water) (Table 10).   The
daily intake could still be increased in special circumstances where the
major food items (like rice in certain areas of Japan) or water (e.g.
some well water) are polluted by cadmium.

In Summary

For the general population food constitutes the most important source of cadmium.   Under normal circumstances drinking water and ambient air contribute
relatively little to the daily intake.   In certain circumstances, however,
(persons living very close to an emission source, heavy smokers) and because
of a higher absorption rate from the lung than from the gastrointestinal tract
(see next section) airborne cadmium may become an important source of exposure

T A B L E   1 0

TOTAL INTAKE OF CADMIUM (μg/day)

	Rural area		Urban area		Industrial area	
	Non smoker	Smoker**	Non Smoker	Smoker**	Non Smoker	Smoker**
From air* (deposition)	0.0005-0.215	3.0005-3.215	0.01-3.5	3.01-6.5	0.05-25	3.05-28
From food		4 to 84	(median = 43)			
From water		2 to 10	(3)			

*   Assuming a daily inhalation of 20m^3 , 25% deposition

**   Assuming a deposition of 3μg/day from cigarettes (40 cigarettes/day)

V - 2   ABSORPTION

2.1 By Inhalation

The amount of cadmium <u>absorbed</u> through the lung depends on the amount retained (Q deposited - Q rapidly eliminated via the upper respiratory
tract by the clearance mechanisms) and probably also on the chemical form
of the retained particles which determines its rate of solubilization in
biological media.

Princi and Geever (395) have demonstrated that in dogs inhaled cadmium
oxide dusts were more readily absorbed than cadmium sulphide. Cadmium
chloride is also more easily absorbed through the lung than cadmium sulphide (390). We have no precise human data to evaluate the quantity of
cadmium deposited in the respiratory tract which is available for absorption and the rate of this process. Lewis et al (273) have found that in
male non smokers (±60 years old) the mean concentration of cadmium in
kidney, liver and lung was 6.6mg and in persons who have smoked the
equivalent of 1 packet of cigarettes per day for 40 years 14mg (see
section 3 of this Chapter). If the cadmium content of kidney + liver +
lung represents 50% of the total body burden (see section 3 of this
Chapter) this means that smokers have accumulated ±14.0mg more than non
smokers. The amount retained per day due to smoking would then be

$$\frac{14000\mu g}{40 \times 365} = 0.96\mu g$$

Neglecting the excretion (which is a small fraction of the amount absorbed) and assuming that the cigarette smoke inhaled each day contains
3μg cadmium the fraction absorbed would then be $\frac{0.96}{3}$ x 100 = 32%. If
the amount deposited in the alveolar compartment is 50% of the amount
inhaled (particle size <0.3μ) one can conclude that approximately 64% of
the amount deposited is absorbed. The true absorption rate is probably
higher since excretion was not taken into consideration. If one makes
the same calculation but assuming that approximately 2μg cadmium are
inhaled per packet then one arrives at about 96% absorption of the amount
deposited.

In rats Moore et al (315) found a retention of approximately 40% in whole body (not counting the gut content) after exposure to an aerosol of $CdCl_2$ (mean particle size: 0.5μ). Since about 10% of the amount inhaled was found in the lungs after exposure the absorption rate of the amount deposited in the lung must have exceeded 75%. An absorption rate of 64% of the amount deposited seems therefore reasonable.

Assuming that 64% of the cadmium deposited can be absorbed that means that in the general environment 13.0-19.0% of the cadmium inhaled is absorbed. This estimate agrees well with the results obtained by Potts et al (390) in mice using a radioactive cadmium chloride aerosol (particle size <2μ). This absorption rate (64%) does not necessarily apply to industrial situations where the particle size distribution of airborne cadmium is much different than in ambient air. It should also be pointed out that after deposition in the respiratory tract some particles may be transferred to the gastrointestinal tract where a fraction may also be absorbed.

2.2 By the Gastrointestinal Tract

Experiments on 5 human volunteers (19 to 50 years old) who were given labelled Cd orally indicate that the absorption rate ranges between 4.7 and 7% (403). This estimate is in agreement with calculations based on balance studies in humans (129).

In animals absorption rates between 0.5 and 12% (on the average 2%) have been reported according to the animal species and the chemical form of cadmium (74, 81, 129, 213, 315). In animals, low calcium diet, iron deficiency and protein deficiency, can stimulate cadmium absorption (on the average by a factor of 2) (162, 238, 251, 475). A calcium deficient diet increases also the retention of cadmium administered parenterally to hamsters (310). Stowe et al (467) have found that in rats the dietary concentration of pyridoxine (vitamin B6) will influence cadmium absorption. Matsusaka et al (cited by Nordberg 1975 (345)) have found that the absorption rate of cadmium in young mice was higher than in adults. Two weeks after an oral dose of radioactive cadmium the whole body retention was about 10 and 1% respectively. Thus, age may play a role in the rate of gastrointestinal absorption of cadmium in man.

2.3 Skin Absorption

Skog and Walberg (451) applied Cd labelled $CdCl_2$ to guinea pig skin and found 1.8% absorbed in 5 hours.   Since the probability of skin contact is low it appears that absorption of Cd through the skin is, in practice, insignificant.

2.4 Total Amount Absorbed

The amount of cadmium absorbed daily for a European or American adult non occupationally exposed is summarized in Table 11.

If we except special circumstances like living very close to a Cd emission source it appears according to these estimates that food and heavy smoking constitute the largest sources of assimilation.

V - 3   TRANSPORT, DISTRIBUTION AND BODY BURDEN

In human beings not excessively exposed to cadmium, the normal cadmium concentration in blood appears to be below 1µg/100ml whole blood but there are large discrepancies between reported mean or median values (0.06-15.9µg/100ml) (41, 51, 84, 85, 96, 129, 246, 287, 348, 415, 481, 510, 530, 543).

In the industrial area of Stolberg in Germany an average cadmium concentration of 0.06-0.11µg Cd/100ml blood was found in two groups of children.   The average Cd concentration in a group of adults from the same region was 0.11µg cadmium per 100 ml blood (313).

A recently determined whole blood cadmium concentration in 11 year old children living near a smelter and in a rural area obtained the following results: children living at less than 1 km from the smelter: 0.11µg/100ml; children living at more than 2 km from the smelter: 0.06µg/100ml; children in rural area: 0.07µg/100ml (Roels, Buchet, Lauwerys, Bruaux: unpublished results).

T A B L E   1 1

Total amount of Cd absorbed (in μg/day)

	Rural area		Urban area		Industrial area	
	Non Smoker	Smoker**	Non Smoker	Smcker**	Non Smoker	Smoker**
From air* :	0.00032-0.1376	1.92032-2.0576	0.0064-2.24	1.9264-4.16	0.032-16	1.952-17.92
From food*** :	range = 0.24 - 5.04		(median = 2.58)			
From water :	range = <0.12 - >0.6		(median = 0.18)			
Total	0.36 - 5.78	2.28 - 7.70	0.37 - 7.88	2.29 - 9.8	0.39 - 21.64	2.31 - 23.56

*   assuming a daily inhalation of 20 m³, 25% deposition, 64% absorption

**   assuming a deposition of 3μg/day from cigarettes, 64% absorption

***   assuming 6% absorption

N.B.   These figures should not be considered as precise because they represent estimates derived from the previous data.

Smokers have a higher blood cadmium concentration than non smokers (51, 97, 252, 506, 530). According to Szadkowski (480) in normal persons the ratio of serum and red cells concentration is 0.5. Falchuck et al (108) found 0.17 and 0.22μg Cd/100ml in two normal human sera and according to Wilden (530) only 10% of blood cadmium is in plasma. Shaikh and Lucis (441) found that Cd is rapidly cleared from rat plasma following injection of radioactive cadmium. Most of the cadmium in blood cells (±60%) is probably bound with metallothionein, a metal binding protein of low molecular weight (342). Haemoglobin could also be a binding protein for cadmium (46, 339). There is not yet any definite indication whether cadmium in blood reflects exposure or body burden. The analytical difficulty for accurate determination in blood must be solved first before answering this question (see Chapter IV). In cadmium workers cadmium level in blood is higher than in the general population. In 19 workers exposed to cadmium, 13% of cadmium was found in plasma compared with 31% of cadmium in 22 control workers (553).

In workers, Piscator (cited by Friberg et al (129)) and Lauwerys et al (256) found no correlation between cadmium in blood and exposure time. Kjellström (personal communication) has found that in newly exposed workers Cd in blood increases without concomitant change in cadmium in urine. All these results would suggest that cadmium in blood is probably not a reflection of body burden but is mainly influenced by current exposure.

Available data from animal experiments does not throw any light on this question. Further research is certainly necessary to confirm this tentative conclusion but as indicated above the analytical difficulties encountered in the analysis of cadmium in blood must first be overcome (255).

Analysis of newborn tissues by Henke et al (175) suggests that at birth the human organism contains virtually no cadmium (less than 1μg). Results obtained in Japan are different since it has been reported that in 14 second trimesters, in the foetuses derived from women who did not live in polluted areas cadmium was present in 80% of livers (mean: 0.113μg/g wet weight), 28% of kidney (mean: 0.05μg/g) and 17% of brain specimens (mean 0.140μg/g) (52).

Flick et al (120) cite an average value of 0.05μg/g in human fetus. Cadmium accumulates with age (at least until age 50, see below) and about 50% of the accumulated cadmium is found in the kidney and the liver. In these tissues, cadmium is probably bound mainly to metallothiomein, a protein of low molecular weight (10.000-12.000) very rich in cysteine residues and deficient in aromatic amino acids (207, 291). Metallothionein has been detected in human kidney, liver, heart, brain, testis, skin epithelial cells and in human embryonic fibroblasts from skin, muscles and lung (42, 283, 402, 416, 477, 535). In animals, the protein has also been found in placenta, spleen and intestinal mucosa (6, 536). Metals other than cadmium can bind metallothionein in vivo i.e. Zn, Cu, Hg, Ag, Sn (417). However no in vivo binding of Pb, Ni, Ir, V, Te, As, Mn, Se, Co, Mo and Bi to metallothionein was observed in rat liver by Webb (520) and Sabbioni and Marafante (417). It has been suggested that this protein plays a role in zinc metabolism (42) and perhaps also in copper metabolism (104). According to Webb and Stoddart (524) two forms of a single Cd^{2+} binding protein exist in the rat liver and these can interact very strongly with a glycosaminoglycan to form a number of adducts with different isoelectric points or molecular weights.

Examining autopsy specimens from 40 patients Syversen (477) has found that the cadmium binding protein in liver and kidney from 19 patients and the mean cadmium content of kidney in this group was higher than the corresponding value for all 40 patients.

The synthesis of metallothionein in liver, kidney and intestine may be induced by cadmium (70, 376, 377, 417, 418, 460, 471, 520, 534, 535) and results obtained by Sabbioni and Marifante (417) suggest that only cadmium (and no other metals) acts as highly specific inducer of metallothionein. Piscator first suggested that toxicity occurs when metallothionein available is insufficient to bind all the cadmium (129). One has also hypothesized that metallothionein is involved in cadmium absorption and transport. Nordberg (342, 345) has found that in mice injection of cadmium bound to metallothionein is more toxic to the kidney tubule than cadmium alone which suggests that metallothionein plays a role in the elicitation of renal tubular impairment at chronic exposure (glomerular filtration followed by tubular reabsorption).

This theory is ir agreement with previous observations that various thiols administered at the same time, or 24 hours after, cadmium, increase its concentration in kidney of mice (155, 351). Gunn et al (155) have previously demonstrated that cysteine increases the kidney toxicity of cadmium in mice but protects the testes. Recently Nordberg et al (345) have demonstrated tnat metallothionein bound cadmium injected intravenously to mice concentrated more in the kidney than free cadmium. This has been confirmed by other authors in female rats (57, 58). After single intravenous administration of labelled $CdCl_2$ (200µg), only a small amount of ^{109}Cd accumulated in the kidney (1.5%) as compared to liver (51.7%). However, cadmium bound to metallothionein accumulated more in kidney than in liver at all dose levels (10 to 200µg cadmi per animal). The urinary excretion of cadmium after administration of $CdCl_2$ was not significant (0.003% of the administered dose). However, in rats injected with cadmium (200µg) as metallothionein, a considerable amount of ^{109}Cd was excreted in urine (31.9%). Cherian and Shaikh (57) also demonstrated that intravenously injected metallothionein is partly degraded in kidney within three hours. The cadmium released from the breakdown of metallothionein is sequestered by the higher molecular weight proteins in the kidney.

Cherian and Goyer (58) found also that intraperitoneal injection of isolated cadmium metallothionein complex (1.2mg Cd/kg) results in accumulation of a large percentage of cadmium in kidney (about 60-70%) and in changes in the ultrastructure of renal tubular cells - thus confirming previous reports - but these morphological results were in contrast to repeated injection of cadmium chloride (0.6mg Cd/kg) to rats 5 days for 4 weeks. There was no damage to renal cells in latter experiments although kidney concentrations of cadmium in these rats are much higher than that in cadmium-metallothionein injected rats.

These results suggest that the cadmium-metallothionein complex is a nephrotoxin when injected to rats but when it is synthesized within the cell it may protect from cadmium toxicity temporarily. Metallothionein synthesis can also be induced in human skin epithelial cells in vitro and then the cells become resistant to the toxic effect of cadmium (416). Since some differences have been observed between metallothionein from the liver and the kidney (522) and since metallothionein can be induced in kidney cells in vitro (523) it has been suggested that metallothionein bound cadmium in the kidney does not necessarily result from redistribution of cadmium from liver to kidney. Furthermore the results of Cherian and Shaikh (57) in female rats demonstrating a rapid clearance of metallothionein from the blood and significant excretion of the protein in urine after its intravenous administration do not indicate a transport role for metallothionein. The protective role of metallothionein against the acute toxicity of cadmium has also been recently ques-

tioned (see Chapter VI).   Goyer et al (557) found in rats that the appearance
of cadmium-metallothionein complex in plasma may correlate with the onset of
irreversible cadmium-induced nepuropathy.

In the U.S. at age 50 the body burden of non occupationally exposed persons
has been estimated by Schroeder et al (432) at approximately 30mg of which 33%
is found in the kidney, 13.8% in the liver, 2.3% in the lung, 0.3% in the
pancreas and the remainder in other tissues.   In 11 subjects from New Zealand
not occupationally exposed to Cd (mean age 39 years - smoking history unknown)
Johnson (204) found a mean Cd concentration in liver of 4.5 ppm (range 1-26.7
ppm).   In England analyses of liver and kidney samples of autopsy material
from 20 persons not occupationally exposed to cadmium give mean values of 2.03
ppm in liver and 11.7 ppm in kidney (77).   In 91 patients from Brisbane
(Australia) with ages ranging from six months to 93 years, the mean cadmium
concentration in kidney was 16.8µg/g whole kidney with a range of 1.0-40.3µg/g
whole kidney (wet weight) (312).   The maximum value was found in the 50 year
old group (mean level in whole kidney: 30µg/g wet weight).

Piscator and Lind (384) and Ostergaard and Clausen (356) who determined cadmium concentration in kidney cortex from Swedish and Danish patients arrived
at the same conclusion that Cd level increases progressively with age until
age 50 where its level reaches about 28-33µg/g wet weight.   Ostergaard and
Clausen (356) did not find difference in the cadmium content of kidney between
patients from cities and those from provinces (15 subjects in each group).

Hammer et al (164) found also that average renal cadmium levels increased over
ten fold from infancy to middle age and then declined somewhat thereafter.
In 14 subjects with mean age of 2 years and in 13 subjects with mean age of 55
years the mean cadmium concentration in renal cortex was 2.6 and 37.75µg/g
wet weight respectively.   The mean cadmium to zinc ratio in renal cortex
increases from 0.06 at 2 years to 0.76 at 55 years and this change in the
cadmium to zinc ratio was primarily due to the marked changes of renal cadmium
concentration with age.   The same authors found that hepatic cadmium increased 4-5 fold from infancy (0.62µg/g wet weight) to adulthood (±3.3µg/g).
Patients with different types of cancer had higher and more variable renal
cadmium levels than non cancer patients.

Sumimo et al (472) analysed tissues from 30 Japanese patients (15 males and 15 females, average age 39 years). They found that the total body burden was higher than 35 mg. The total cadmium content (in mg) of various organs was: muscle: 7.0; bone: 0.82; fat: 0.45; blood: 0.76; skin: 1.3; liver: 8.5; brain: 0.16; digestive tract: 0.75; lung: 0.65; heart: 0.048; kidney: 12; spleen: 0.12; pancreas: 0.27. They found a statistically significant correlation (r = 0.96) between total mercury content and cadmium content of tissues. Cadmium accumulation was higher in females, even when tissue weights were taken into account.

Smokers have higher concentration of cadmium in tissues (kidney, lung, liver) than non smokers (164, 180, 273, 302). Lewis et al (274) determined cadmium concentrations in kidney, liver and lung derived from 172 American adults (mean age 60 years). The mean values for total organ content of the metal were: non smokers: 4.16mg for kidney, 2.28mg for liver, 0.36mg for lungs; smokers: 10.28mg for kidney, 3.06mg for liver and 0.81mg for lungs. They estimated that adult American non smokers (mean age 60 years) have on average a total body burden of about 13mg cadmium (based on the assumption that composite kidney, liver and lung values constitute 45-50% of this total). On the other hand the adult cigarette smokers (mean age 60 years) have a total body burden of 30mg. Hammer et al (164) found approximately the same values. They calculated that non smoking males aged 40 to 70 would have about 5.2mg of cadmium in their kidneys and about 3.7mg of cadmium in their liver for a combined total of 8.9mg: comparable figures for males smoking half a packet or more daily would be 11.4mg in the kidney and 7.5mg in the liver for a combined total of 18.9mg of cadmium in these two organs. Assuming the amount in the kidney and liver to represent approximately 50% of the adult body burden of cadmium they estimated this to be about 17.8mg of cadmium for male non smokers and about 37.8mg of cadmium in smokers. Applying the same reasoning to data published by Shuman et al (448) the average body burden for adult non smokers would be 19.2mg and for smokers 32.4mg. The American data is thus in good agreement, and the difference between non smokers and smokers suggests that more than half of the cadmium found in the latter group was due to accumulation through cigarette smoking. Friberg et al (129) have estimated the total body burden of a 50 year old adult at approximately 15-20mg in U.K. and Sweden and 80mg in non polluted areas of Japan. From a more recent study performed in Sweden it was calculated that the total body burdens for non-smokers increased continuously from age 25 to age 65 when it reached a value of 9.5mg.

Kidney and liver together contained about 77%, 41% and 38% of the cadmium body burden at ages 25, 55 and 65 respectively (Kjellstrom). This confirms that the daily cadmium intake in Sweden is lower than in most other countries. Results obtained by Sumimo et al (472) indicate that in Hyogo Prefecture (central part of Japan) the adult body burden is probably lower (±40mg) than Friberg's estimate. In not excessively exposed individuals the concentration of cadmium in kidney is 2.6 to 43 times higher than in liver (129). In Cd exposed individuals, however, this ratio is lower and liver concentration may even exceed kidney concentration (455). The total body burden of cadmium in workers could exceed 1200mg (129).

Cadmium concentration is higher in renal cortex than in the medulla (about 1.5 times higher). Livingston (278) has observed that, in kidney, zinc and cadmium show a concentration gradient with the highest concentration in the outer surface of the cortex.

Contrary to Sumimo et al (see above), Anke and Schneider (9) and Hammer et al (154) found that average renal cadmium concentrations in males generally exceeded those of females. This could partly reflect different smoking habits since smokers have higher concentrations of cadmium in tissue (kidney, lung, liver) than non smokers (180, 302). In normal human renal cortex there is an equimolar increase in zinc content with increasing cadmium levels up to the age of 50 (384). After this age, there is a lowering of the cadmium-zinc ratio (384, 432).

In horses Piscator (385) found that at low renal cortex cadmium concentration (<60µg/g wet weight) the increase in zinc was equimolar to the increase of cadmium. However, with increasing concentration of cadmium, zinc did not increase to the same extent, thus indicating a disturbance in the cadmium-zinc relationship.

Schroeder et al (432), Piscator and Lind (384) and Miller et al (312) have found that the cadmium content of kidney decreases after age 50 when mean concentrations in the renal cortex in persons from different countries have been shown to range from 20-60µg/g wet weight. Higher values have been shown in certain areas of Japan (up to 125µg/g wet weight) and in exposed workers (up to 300µg/g wet weight).

It is, however, possible to find lower cadmium concentrations in kidney of exposed workers and Itai Itai patients than in normal individuals because when
renal damage is present, urinary cadmium excretion is increased (411) and
therefore renal level may decrease (195). Several hypotheses have been put
forward to explain the decreased cadmium concentration in kidney cortex with
age found in normal individuals.

1) Increased exposure of the population to cadmium over the last 50 years
   (129). This hypothesis is consistent with the fact that 70% of the
   total world production of cadmium has occurred within the last 20 years.

2) Increased mortality for human beings with higher renal levels (432).

3) Failure to control for sex, smoking status, diseases at death; renal
   changes with age and similar covariates could explain some of the findings. For example some studies may have included a larger proportion of
   smokers or patients with cancers in the 40-60 year old age group which
   would increase renal cadmium levels in this age group (164).

4) Increased renal cadmium loss with age (164).

5) Decreased accumulation due to diminution of smoking habits in older age
   group might also account in the past, for the decline with age (164).

The first hypothesis seems to be the most plausible. Cadmium concentration
in pancreas increased also markedly in exposed persons (from 1-65µg/g wet
weight). In bone cadmium level is usually low (below 1µg/g) (129, 472).
Animal and human data indicate that cadmium does not easily penetrate into the
brain (472, 483).

V - 4   EXCRETION (of absorbed cadmium)

4.1 Urinary excretion

Cadmium is mainly excreted via the urine. In normal adults the amount
excreted daily via the urine is probably below 2µg/day (for a tabulation
of values reported in the literature see reference 129: range of means
0.2-3.1µg/l). In cadmium exposed workers urinary excretion can be much
higher (up to several 100µg/day) (129). Whether cadmium concentration
in urine reflects body burden or current exposure is still debatable.
We have summarized below results suggesting either hypothesis (256).

A. Reports suggesting that urinary cadmium excretion does not reflect body
    burden (but may reflect exposure).

1. Data published by Szadkowski et al (479) in Germany suggest that
    there is practically no increase of urinary cadmium excretion with
    age among "normal" persons. Recently, however, Szadkowski (481)
    has stated that there is a "pronounced" dependence of urinary cad-
    mium concentration on age.

2. Suzuki and Taguchi (476) found no correlation between cadmium in
    urine and age in Japan.

3. In more than 100 workers highly exposed to cadmium but without
    kidney lesion, Lauwerys et al (256) have found no correlation be-
    tween cadmium excretion and length of exposure (assumed to be an
    index of kidney accumulation).

4. Lehnert et al have made the same observation on a more limited
    number of workers - 18 (261).

5. Piscator, who examined 20 Swedish workers exposed to cadmium in an
    alkaline battery factory, reached the same conclusion. His data
    has not yet been published (cited by Task Group on metal accumula-
    tion 1973) (483).

6. In Japan, Harada seems to have found a correlation between exposure
    level and cadmium concentration in urine (personal communication
    from Tsuchiya). Watanabe and Murayama (518) made the same observa-
    tion but in adults only. The Ministry of Health and Welfare of
    Japan concluded on the basis of various epidemiological studies that
    there is a direct relationship between daily oral intake and the
    amount of cadmium excreted with urine daily.

    $$x = 33.9y + 2.5$$

    where x = amount of cadmium excreted with urine daily and
          y = cadmium intake in mg/day.

    A daily oral intake of 300µg would thus correspond to a urinary cad-
    mium excretion of 13µg/24h (349). In Japanese adults, Watanabe and
    Muryama (518) found the following relationship between cadmium con-
    centration in unpolished rice and cadmium concentration in urine:

Cadmium in rice (ppm)	Cadmium in urine (µg/l)
0.33	7.1
0.35	7.7
0.61	11.1
0.88	13.0
1.10	14.2

7. In workers after exposure has ceased Adams et al (3) found that cadmium excretion decreases.

8. In adults living in cadmium polluted areas of Japan, Watanabe and Murayama (518) found no increase in cadmium in urine with age.

Age groups	Cadmium in urine	
	Low polluted area	High polluted area
30 - 39	7.75	11.50
40 - 49	8.45	13.10
50 - 59	6.00	14.04
60 - 69	6.22	12.44
70 - 79	10.29	11.61
80 - 89	8.33	11.51

9. In workers exposed to cadmium, Materne et al (297) found a low but statistically significant correlation between cadmium levels in blood and in urine ($r = +0.55$). Since it is believed that blood concentration reflects more recent exposure than body burden their results suggest that in case of high exposure cadmium in urine is mostly affected by the magnitude of this exposure.

10. Watanabe and Murayama (518) found about the same correlation coefficient among Japanese adults exposed to environmental pollution by cadmium ($r = 0.41-0.49$).

B. Results suggesting that urinary cadmium excretion reflects body burden

1. In Japan Katagiri et al, 1971 (cited by Friberg et al, 1974) (129) found the following distribution of cadmium concentrations in urine with age:

Age group	Mean cadmium concentration in urine ($\mu g/l$)
4 - 6	0.47
9 - 10	0.65
14 - 15	0.72
20 - 29	0.99
30 - 39	1.13
40 - 49	1.76
50 - 59	1.75

2. Tsuchiya et al, 1974 (personal communication) and Shiroishi et al (447) found also the same trend. It should, however, be stressed that the scatter of the results is quite high. In each age group the standard deviation of the results has approximately the same value as the mean.

3. In male adults working in an environment slightly polluted by cadmium (control workers working in the same factories as cadmium exposed workers), Lauwerys et al (256) found a low but nevertheless statistically significant correlation between length of employment and cadmium concentration in urine.

4. In workers newly exposed to cadmium, Kjellström found no increase in urinary cadmium for one year (personal communication). In 1967 already Tsuchiya (504) reported the absence of increased cadmium concentration in urine in workers exposed to cadmium for less than a year.

5. In polluted areas of Japan a positive correlation was found between cadmium concentration in rice and cadmium concentration in urine of adults but no correlation was found in children (517, 518).

6. Studies performed by Nordberg (343) on mice (repeated subcutaneous injections of cadmium) suggest that on a group basis only urinary excretion of cadmium is correlated with body burden (in the absence of tubular damage).

7. Nomiyama et al (335, 338) administered cadmium-chloride subcutaneously to rabbits at a dose of 1.5 mg/Kg/day up to 5 weeks. A significant correlation was found between cadmium level of the renal cortex and urinary excretion of cadmium.

It is, of course, difficult to try to reconcile these apparent contradictory results. The following hypothesis could, however, be formulated (256). At low exposure level (general environmental conditions) the amount of cadmium absorbed may be insufficient to saturate all the body binding sites (e.g. induced metallothionein) and urinary excretion does not increase proportionally to the exposure levels. In these circumstances urinary concentration is principally in equilibrium with body burden. In high exposure conditions (e.g. workers, adult Itai-Itai patients) and in the absence of renal lesion the urinary concentration could be more a reflection of exposure levels (all the binding sites being now saturated). In newly exposed persons a latent period may thus be observed before cadmium in urine correlates with exposure. Anyway, currently because of the large scatter of the results the correlation between cadmium in urine and exposure or body burden are only valid on a group basis. Further research work is urgently needed to clarify all these points. It seems now well documented that when cadmium induced kidney damage is present, urinary cadmium excretion is much increased (3, 253, 197, 411, 553).

Considering the urine as the main excretory route for cadmium and assuming that this excretion is related to body burden Friberg et al (129) have calculated that approximately 0.004 to 0.015% of the body burden is excreted daily which corresponds to a biological half-life of 13 to 47 years.

The determination of urine cadmium concentration in 11 year old children living near a lead smelter whose cadmium level in blood has been reported above and has been compared with the results of those found in matched children living in the rural area. The cadmium concentrations were: 1.06, 0.76 and 0.23µg/g creatinine in children living at less than 1km from the smelter, in children living at more than 2km from the smelter and in those living in the rural area, respectively (Roels, Buchet, Lauwerys, Bruaux, unpublished results).

Rosmanith et al (415) reported rather similar results in Germany. In 400 children from an industrial area with a high degree of lead, zinc and cadmium pollution the average urinary cadmium concentration was 2.42µg/l (SD = 3.02). The cadmium concentration in the urine increased with increasing zinc fall-out.

4.2 Other routes of excretion

In man, fecal excretion (after absorption) appears less important than urinary excretion although its exact evaluation is difficult (it is difficult to determine how much cadmium comes directly from food without being absorbed and how much is really excreted). A study by Rahola et al (403) with radioactive cadmium suggests that in man less than 0.1% of absorbed cadmium is excreted via the feces. Nordberg (342) has found, however, that in mice receiving repeated sub-cutaneous injections of cadmium for half a year fecal excretion is more important than urinary excretion but fecal excretion did not increase at all directly in proportion to the body burden of cadmium. Ogawa et al (351) made the same observation after single intraperitoneal administration of $CdCl_2$ to mice.

In rats biliary excretion of cadmium has been demonstrated (50, 68). Vorstal and Cherian (515) have noted that $^{109}CdCl_2$ given intravenously in rats is excreted in the bile in proportion to the dose. When low doses of cadmium were administered (0.1mg Cd/kg intravenously) only a fraction of 1% of administered dose appeared in the bile but more than 15% of the administered dose appears in the bile after injection of 2mg/kg. Under these conditions the biliary cadmium is in the form of glutathione complex, and cadmium in liver and kidney supernatants is bound to high molecular weight proteins (less than 10% bound to low molecular weight protein).

When cadmium binding protein (metallothionein) was induced by injection of cadmium 24h previously, liver retention increased and biliary excretion decreased (56). In this experiment more than 90% of liver cadmium and 70% of kidney cadmium were bound to a low molecular weight protein in kidney and liver supernatant.

Cikrt and Tichy (68) followed the biliary excretion of cadmium and its excretion through the wall of the gastrointestinal tract in rats after single intravenous administration of $CdCl_2$ in doses of either 67, 90 or 120μg of Cd^{2+} in rats. The cumulative biliary excretion reached 24 hours after administration of the 67μg dose was 0.83%, at 90μg 1.18% and after the 120μg dose 5.68% of the amount given. The highest excretion rate of Cd^{2+} was detected between 15 and 30 minutes after administration. Chromatographic fractionation on Sephadex of bile indicated that cadmium was bound with both the high (>100,000) and the low molecular weight fraction (170).

The mean amount of cadmium found in the content of the entire gastrointestinal tract and feces was 5.5% of the administered dose. These results are in agreement with those previously obtained by Decker et al (1957) who found 7.3% of cadmium in the feces 24 hours after the administration of a single intravenous dose of ^{115}Cd (0.63mg/kg) to rats. Tsuchiya et al have found cadmium in human bile and have concluded that there is an entero-hepatic circulation of the metal (cited by Friberg et al (129)).

Since, in humans, excretion of cadmium is apparently low, cadmium concentration in feces is mainly an indicator of the amount of cadmium ingested. The amount of cadmium excreted via skin, hair, sweat, saliva and milk is considered of minor importance in comparison with that excreted via urine and gastrointestinal tract (for a review, see ref. 129).

In saliva, Dreizen et al (92) found concentrations below 0.1μg/g. Animal data indicate that although cadmium is taken up in the mammary gland and can reduce milk production it is excreted only in small amounts via milk (284, 309). In hair, mean concentrations ranging from 0.54 to 3.5μg/g have been reported (129, 163, 436). Rosmanith et al (415) found an average cadmium concentration in hair of 1.76μg/g (S.D.: 1.73) in 400 children from an industrial area in Germany.

In children Hammer et al (163) found a good correlation between ambient cadmium concentration and hair levels. Truhaut and Boudène (502) were the first to propose evaluating the degree of exposure of workers to cadmium by determining its concentration in hair.

Examining 50 autopsy specimens Oleru (353) found that cadmium concentration in hair was significantly correlated with cadmium concentration in kidney (r = 0.52) and in liver (r = 0.36) but not with cadmium concentration in lung (r = 0.15).

In mice, Nordberg and Nishiyama (343) found a close correlation between whole body and hair concentration of cadmium following an intravenous injection of radioactive cadmium. On the other hand, Vuori et al (516) who analysed samples from lung, muscle, liver, kidney and hair from 20 accidentally deceased persons could find no correlation between cadmium level in hair and other tissues. Watanabe and Murayama (518) found also no correlation between cadmium in hair and cadmium pollution. Furthermore, Nishiyama and Nordberg (331) have demonstrated that it is highly difficult to distinguish between endogenous cadmium and cadmium externally deposited on the hair.

Conflicting reports have been published regarding the placental transfer of cadmium in animals (20, 114, 284) but this may be due to differences in various experimental factors (e.g. gestational age, species, and strain differences (536)). Lauwerys et al have measured cadmium concentration in blood of 106 mothers living in Belgium and of their new-

borns (unpublished results). The average blood concentration of cadmium
was: mother = 0.25µg/100ml whole blood; newborns = 0.21µg/100ml whole
blood. In the placenta the mean cadmium concentration was 1.64 µg/100g.
Baglan et al (18) analysed approximately 100 placentas, maternal and
fetal bloods from persons living in Tennessee and found the following
results for Cd: placenta: 1.7µg/100g (wet weight), maternal blood:
1.69µg/100g, fetal blood: 1.62µg/100g. The average level found by
Dawson et al (1968) for placenta samples obtained near Galveston, Texas,
was 800µg/100g. An analytical error is certainly responsible for such
a high value. Thürauf et al (497) measured the cadmium content of 179
placentas from 3 different regions of F.R. Germany. The average content
was 60µg/placenta (i.e. 9.2µg/100g). A significant difference was
observed between the mean Cd content in the placentas coming from the
Rhine-Ruhr area and those coming from the Bavarian forest. The data of
Lauwerys and others from the literature (see ref. 483) indicate that
cadmium can cross the placenta in humans. Of course the body burden of
the newborn will remain low since exposure occurs only during 9 months to
a very low level (cadmium concentration in mother's blood).

Chapter VI

TOXICOLOGICAL EFFECTS
OF CADMIUM

Cadmium has no known biological function and is highly toxic to mammals.

In this review special attention has been given to the health risks for the general population which are associated with long term exposure to cadmium. Before dealing with this aspect, the potential hazards resulting from acute or short term cadmium exposure have been briefly reviewed.

For each toxicological action is discussed the information, when available, derived from human observations and from experimental investigations. The information available regarding the carcinogenicity, mutagenicity and embryotoxicity of cadmium is dealt with separately.

VI - 1   ACUTE AND SHORT TERM EXPOSURE

In human beings the two main target organs of cadmium are the gastrointestinal tract after acute ingestion and the lungs after inhalation. The lesions induced on these sites have been considered first. The other acute effects which have been observed mainly if not exclusively in animals have then been reviewed.

1.1 Oral administration (target organ in human beings: gastrointestinal tract)

The acute oral LD_{50}s of cadmium compounds for various animal species are presented in Table 12. Depending on the chemical form the oral lethal dose 50 (LD_{50}) in rats varies from >5000 to 16mg/kg.

T A B L E   1 2

Acute LD_{50} of cadmium compounds

Substance	Species	Route	LD_{50} (mg/kg)
Cadmium acetate	rat	p.o.	333
Cadmium bromide	rat	p.o.	322
Cadmium chloride	rat	p.o.	88
	rat	p.o.	302
	guinea-pig	p.o	63
	mouse	p.o.	5
	mouse	p.o.	175
Cadmium carbonate	rat	p.o.	659
	rat	p.o.	438
Cadmium cyanide	rat	p.o.	16
Cadmium fluoroborate	rat	p.o.	250
Cadmium fluoride	guinea-pig	p.o.	150
Cadmium fluorosilicate	rat	p.o.	100
Cadmium formate	rat	p.o.	162
Cadmium iodide	rat	p.o.	222
Cadmium nitrate	rat	p.o.	397
Cadmium oxide	rat	p.o.	72
	rat	p.o.	123
	rat	p.o.	130
	rat	p.o.	296
Cadmium perchlorate	rat	p.o.	875
Cadmium (metal) powder	rat	p.o.	2330
Cadmium propionate	rat	p.o.	300
Cadmium selenide	rat	p.o.	>5000
Cadmium stearate	rat	p.o.	1225
Cadmium succinate	rat	p.o.	660
Cadmium sulphate	rat	p.o.	357
Cadmium sulphide	rat	p.o.	>5000
Cadmium acetate	rat	I.P.	74.9

T A B L E   12 (continued)

Substance	Species	Route	LD_{50} (mg/kg)
Cadmium chloride	rat	I.M.	25
	rabbit	I.V.	2.5-3.0
	mouse	I.P.	4.3
	mouse	I.P.	<6.5
Cadmium oxide	mouse	S.C.	94
	rat	I.V.	25
Cadmium succinate	mouse	I.P.	270
Cadmium lactate	mouse	S.C.	13.9

From references: 8, 67, 106, 244, 280, 304, 503, 544.

In man, acute oral intoxication usually results from the ingestion of food or beverages which have been contaminated with cadmium usually during storage in cadmium plated containers (107). The first symptoms which occur after acute ingestion of cadmium are severe nausea, vomiting, diarrhea, muscular cramps and salivation (11). In the case of fatal intoxication these symptoms are followed either by shock due to the loss of liquid and death within 24 hours or by acute renal failure and cardio-pulmonary depression and death within 7 or 14 days. Liver damage may also be observed (353).

The prohibition by the health authorities of the use of cadmium in contact with food and drink has much reduced the incidence of this form of acute poisoning. An estimate of the acutely toxic oral doses of cadmium for man can be made from accidental cases of poisoning.

Acute poisoning has resulted from consumption of coffee and wine containing 800 and 126 ppm cadmium respectively (503). Punch containing 67ppm cadmium has caused sickness. If it is assumed that 200-500ml were consumed this represents a cadmium intake in the range of 13-35mg (300). Cases of acute poisoning in school children have been reported from Sweden (129). The children had consumed a fruit drink from a "distributing machine" which had a cadmium-plated water reservoir. The concentration of cadmium found in the water used for dilution ranged from 0.5-16ppm .(48).

On the basis of various literature reports a rough scale of acute oral toxicity of cadmium to man has been elaborated (183, 511):

Doses	Effects	References
3-90 mg	- emetic threshold	11, 300, 400
	- reported non fatal incidents	465
15 mg	- experimentally induced vomiting	40, 300, 465
10-326 mg	- severe toxic symptoms but not fatal	148, 465
350-3500 mg	- estimated lethal doses	148
1530-8900 mg	- reported lethal doses	300, 535

In summary it can roughly be estimated that 3 mg is the no-effect level of cadmium administered as a single oral dose to man.

1.2 Inhalation (target organ in human beings: respiratory system).

Acute inhalation of freshly generated cadmium fume is a well known hazard in industry (mostly for welders) but since usually there is no or only slight discomfort at the time of exposure a lethal exposure is possible without warning (17, 32, 94, 457, 533, 547). The arc-generated fume is considered twice as toxic as the thermally generated fume.

Symptomatology is that of severe bronchial and pulmonary irritation appearing several hours after exposure. It includes: irritation and dryness of the nose and throat, cough, headache, dizziness, weakness, chills, fever, severe chest pain, breathlessness. Nausea, vomiting and diarrhea may also occur. The delayed-onset pulmonary edema is responsible for the fatal outcome. Rarely hepatonephritis develops (176).

It has been stated that those patients surviving the acute episode recover, apparently, without major permanent damage or disability (461, 500) but human data are too limited to definitely answer the question whether non fatal acute exposure does or does not produce long-term effect in humans. Zavon and Meadows (547) have suggested that coronary occlusion could follow acute exposure to cadmium fume. Symptoms of early acute cadmium fume poisoning and metal fume fever (a benign condition most commonly due to zinc oxide fume) are extremely alike. Since the treatment and prognosis of the two diseases are quite different, confusion between them may have dramatic consequences.

The acute toxicity of cadmium compounds administered by inhalation to various animal species is summarized in Table 13.

T A B L E   1 3

Acute toxicity of cadmium compounds by inhalation

Substance	Species	LC_{50} (mg/m^3)
Cadmium oxide	rat	500-1300 mg.min/m^3
	mouse	700 (10 minutes)
	mouse	250 (120 minutes)
	dog	4000 (10 minutes)
	rabbit	2500 (10 minutes)
	guinea-pig	3500 (10 minutes)
	monkeys	1500 (10 minutes)
Cadmium stearate	rat	130 (2 hours)
Cadmium chloride	dog	LC_{90} = 8 (30 minutes)

From references 23, 24, 67, 167, 244, 421.

The biochemical damage found in rat lung exposed for 2 hours to an aerosol concentration of 10mg of $CdCl_2$ per m^3 are rather similar to those found following exposure to known lung irritants like O_3 and NO_x (172).

Two groups of authors (23, 32) have calculated the lethal dose for man. Their estimations are based on the amount of cadmium found in the lung in fatal poisoning and on the assumption that the percentage of retention of cadmium oxide fumes is about 10%. The following values were found which are in quite good agreement: 2,500-2,900mg min/m^3 (23, 24) and 2,600mg min/m^3 (32) (N.B. the lethal dose of an airborne chemical may be expressed as the product of the atmospheric concentration in mg/m^3 and the exposure time in minutes. This formulation assumes that the product (concentration time) to produce a certain effect is a constant. This assumption is usually only valid for short exposures and the limitation of such expression has been recently stressed (286). On the basis of the preceding observations, Friberg et al (129) have estimated the leth concentration of cadmium oxide fume for human beings to be around 5mg/n for an eight hour exposure. This is in satisfactory agreement with th estimate made by the American Industrial Hygiene Association: 140-290mg/m^3 for 10 minute exposure (191).

Since in animal species one fourth the LD_{50} can still give rise to pulr nary lesions Friberg et al (129) concluded that an 8h exposure to 1mg/r cadmium oxide fume is still dangerous for man. The State of Californi Department of Public Health (461) reported one fatal case of acute cad mium inhalation and estimated that the worker had been exposed for two and a half hours to at least 1mg/m^3. According to Wilcox (529), 0.5m n^3 in workroom air has been linked with mild cases of pneumonitis.

No human data regarding the acute toxicity of cadmium dust is available

On rabbits, Friberg (127) has found that the LD_{50} of cadmium dust part icles (size smaller than 5μ) was about 8000mg min/m^3 which is approxi mately 3 times the value for cadmium oxide fume (see Table 13). Appl ing this toxicity ratio (cadmium oxide fume/cadmium dust) to man one c estimate that an 8 hour exposure to 3mg/m^3 "respirable" cadmium dust could still lead to acute respiratory symptoms.

1.3 Other acute effects of cadmium

The following effects of cadmium have mainly if not exclusively been in duced in animals.

a. Effect on the kidney

Although the lung lesions are dominant in acute cadmium oxide fume inhalation some kidney lesions have also been described in man (32, 43) and animals (397, 398). It is probable, however, that the kid ney lesions are not due to cadmium per se but to the coexistence of circulatory disturbances. In a fatal case of intoxication follow-

ing the ingestion of 5g cadmium iodide by a 23 year old man, signs
of damage to kidneys (and liver) were also observed (353).   One or
two subcutaneous injections of cadmium chloride (9-10mg/kg) can produce renal tubular damage in rats and in rabbits (109, 122).

b.   Effect on the liver

Like the kidney, the liver accumulates large amounts of cadmium
after acute exposure (see Chapter V).   It is not therefore surprising that morphologic and functional changes have sometimes been
observed in humans and animals exposed to lethal or near lethal
doses of cadmium (8, 185, 341, 397, 535).   Johnson et al (205) have
reported that 3 days after intraperitoneal administration of 2mg/kg
cadmium acetate to male rats the duration of response to subsequent-
ly administered hexobarbital or zoxazolamine was potentiated.   This
effect was due to changes in microsomal enzyme activity in the
liver.   The same observations have been made in mice (161).
Pretreatment of the animals with cadmium induces a tolerance state
to cadmium-induced alteration of drug response and metabolism (409).
However, no modification of hexobarbital metabolism was found in
female rats injected intraperitoneally with 2mg Cd acetate which
indicates a sex-related difference in the ability of cadmium to
alter drug metabolism in rats (178).   Cadmium nitrate, administered
to mice by intraperitoneal injection in amounts above 0.01mg/kg,
caused a significant decrease in liver microsome oxidative demethyl-
ation in vitro (507).

Mego and Cain (304) have studied the effects of cadmium acetate injections on heterolysosome formation and on the digestion of ^{121}I-
labelled formaldehyde treated bovine albumin in these particles from
the kidneys and livers of mice.   Cadmium inhibited proteolysis in
liver particles 2h after intraperitoneal injections at LD_{50} doses
(4.3mg Cd/kg), and the effect became progressively more pronounced
up to 20 hours.   At this time, kidney particles were also affected
but to a lesser degree.   Cadmium decreased also the uptake of labelled protein into subcellular particles, particularly in the liver
at 24h after injection:

Pretreatment of mice with phenobarbital produced a reduction in mortality of male mice injected with cadmium chloride (6.5mg/kg I.P.). This protective effect is associated with an increased cadmium accumulation in the liver and a decreased concentration in the kidney (544).

c.   Effect on the cardiovascular system

Single or a few parenteral administrations of cadmium salts (cadmium sulphate, cadmium citrate ...) have been shown to either decrease (78, 368) or increase (368, 370, 371, 374, 429) blood pressure depending on the species, the doses and the route of administration. The lowest dose reported to still induce an effect is 0.02mg Cd/kg by intravenous injection in rats (370).   In female Long Evans strain, one or two intraperitoneal injections of 1-2mg Cd/kg as cadmium citrate produce a hypertension which, over a month's period became equal to the hypertension appearing in rats with partial constriction of one renal artery (429, 431).   Hyperplasia of the juxtaglomerular apparatus and of the glomerular zone of the adrenal cortex was observed by Chiappino and Baroni (61) after cadmium administration to the Sprague-Dawley rats.   The authors suggest that the induction of the renin-aldosterone system was probably responsible for the hypertension.

d.   Effect on the testes

Acute testicular necrosis may be induced in several animal species (mice, rats, rabbits, monkeys, guinea-pigs, golden hamsters, calves, sheep) after single systemic (subcutaneous, intraperitoneal, intravenous) administration of cadmium salts (45, 88, 145, 157, 201, 234, 358, 359, 360, 364).   Sterility can thus be produced in animals by systemic or intratesticular injection of cadmium.

Some species are resistant, however, to this acute toxic effect of cadmium.   It appears that this effect of cadmium is generally seen in species possessing scrotal testes and is absent in those possessing intra-abdominal testes (203).   Oral administration does not

appear to be very active in this regard. Only very large doses
(20-70mg Cd/kg) administered orally were reported to produce testicular changes in rats (38).   The lowest effective dose that has
been reported to give rise to testicular necrosis is 0.12mg Cd/kg by
the intravenous route in calves (364).

In animals, zinc, selenium, cobalt and thions can antagonize the
necrotizing effect of cadmium (137, 151, 156, 211, 359, 360, 521).
The most efficient protector known is selenium (212, 296).   With
the use of ^{109}Cd it was shown that none of the protective agents
actually lowered the amount of cadmium reaching the testis.   In the
case of selenium there was actually an increased level of cadmium in
the testis (54, 156).

Pretreatment of animals with small doses of $CdCl_2$ reduces also the
necrotizing action of large doses of cadmium (157, 197, 339, 486).
Such a pretreatment makes also the animals resistant to a lethal
dose of cadmium (see below other effects).   It has been suggested
that the protective action is the result of metallothionein induction (440) (see transport, distribution and body burden in Chapter
V).   Nordberg (342) has shown that when cadmium was injected partly
bound to metallothionein no testicular necrosis occurred from doses
of cadmium that gave rise to this effect when injected alone.   On
the other hand, histologically evident tubular kidney damage was present in the metallothionein injected animals but not in the animals
injected with cadmium alone.   Anatomical studies suggest that the
acute testicular damage induced by cadmium might result from a toxic
effect on the vascular system of the testis (21, 63, 152, 289, 295,
439).

A direct action of cadmium on the germinal epithelium has also been
proposed as the mechanism of toxicity (259).   A decrease of carbonic anhydrase activity in rat and mouse testes has been observed
after acute cadmium administration (184, 352), but this enzyme does
not constitute the main target molecule of cadmium since the carbonic anhydrase activity in the testes of fowl, a species resistant to
cadmium is also inhibited by this metal (202).

Omaye and Tappel (354) have found that testicular glutathione peroxidase activity is inhibited following parenteral administration of
cadmium chloride (4.6mg/kg). This inhibition is prevented by prior
administration of selenium which also prevents testicular atrophy
and necrosis. The authors suggest that testicular glutathione
peroxidate may be the direct or indirect target of cadmium induced
testicular necrosis, and this damage results in lipid peroxidation.

Chen et al (53, 54) have recently proposed that a 30,000 molecular
weight Cd-binding protein of the soluble fraction could be the most
likely target of cadmium in the testis. This protein is rather unstable during storage (55).

Testicular damage has never been reported in man acutely exposed to
cadmium.

e.   Effect on the nervous system

Hemorrhagic necrosis in nervous ganglia (gasserian and sensory spinal ganglia) has been produced in rats by subcutaneous injections of
large doses of cadmium chloride (2.5-28mg Cd/kg) (136, 139). Pretreatment with small doses of cadmium or with zinc or glutathion can
prevent the ganglionic damage produced by a single large dose (137,
138). Treatment of newborn rats and rabbits with cadmium causes
cerebrum and cerebellum damage whereas adult animals are resistant
to this effect (138).

No nervous lesion has been described in humans acutely intoxicated
with cadmium.

f.   Effect on the hematopoïetic system

Hamilton and Vallerg (162) found that 10ppm cadmium administered to
mice in drinking water for 8 days reduced the oral absorption of
iron. Lower doses had no significant effect.

g.   Effect on the pancreas

The administration of high doses of cadmium acetate (2.0-6.0mg/kg
intraperitoneally) produces hyperglycemia and glucose intolerance in
mice (142).   The hyperglycemic effect of cadmium is partially mediated through the adrenal gland but cadmium-induced glucose intolerance is the result of an extra-adrenal effect, possibly a decrease
in the secretory activity of the pancreatic betacells (142).
However, recently it has been demonstrated that cadmium enhances
also gluconeogenesis which could be responsible for its hyperglycemic effect (450).   Subacute exposure of rats to cadmium chloride
(1.0mg/kg subcutaneously, twice a day for 7 days) alters several
parameters of carbohydrate metabolism (stimulation of hepatic pyru-
vate carboxylase, phosphoenolpyruvate carboxykinase, fructose 1,6-
diphosphatase and glucose 6 phosphatase, increase of hepatic cyclic
adenosine nonophosphate and circulating blood glucose) and reduces
serum insulin levels.   Administration of selenium dioxide (1.0mg/kg
subcutaneously, twice a day for 7 days) concurrently with cadmium
prevents several of the cadmium-induced metabolic and functional
changes (306).

h.   Other effects

Rats pretreated with Cd^{2+} (1mg/kg intravenously as $CdCl_2$) were
resistant to a subsequent normally lethal dose of the same cation
(4.5 mg/kg intravenously).   Protection induced by the pretreatment
was maintained for at least 3 days and then decreased with time
(525).   Metallothionein was induced by cadmium pretreatment but in
contrast to the resistance against toxicity did not decrease with
time.   Thus these results are not in agreement with the above
mentioned hypothesis that pre-induced metallothionein plays a significant role in the protection against the acute toxicity of cadmium.

Single intravenous administration of cadmium acetate (6mg/kg) enhances the susceptibility of rats to intravenous challenge with E.
coli by approximately 1000-fold (73).   Koller et al (242) have
studied the humoral antibody response in mice after single exposure
to cadmium and found that cadmium caused an increase in IgM antibody

formation and IgG antibody response when injected intraperitoneally
(0.15mg/animal) but resulted in a slight decrease when given orally
(0.15mg/animal).

Conclusions

If in animals acute administration of cadmium can produce toxic effects in
many organs (lung, cardiovascular system, testis, liver, kidney, nervous
system ...) in man mainly local manifestations (gastrointestinal disturbances
and chemical pneumonitis) have been demonstrated to have occurred following
ingestion or inhalation of cadmium.   For a single oral exposure the no-effect
level in man is currently considered to be around 3mg cadmium (per man).   For
an 8h exposure period the no-effect level for cadmium oxide fume and "respirable" dust are probably below 1 and 3mg/m^3 respectively.

VI - 2   LONG TERM EXPOSURE

In man the two main target organs after long term exposure to moderate cadmium
concentrations are the lungs and the kidneys.   Lung lesions result from the
direct local action of inhaled cadmium dust or fume and have been observed
only in workers occupationally exposed to cadmium.   Kidney lesions which
result from chronic absorption of cadmium either through the lung or the
gastrointestinal tract have been described in workers (exposure through inhalation) and in the general population (Itai-Itai disease) in certain areas of
Japan (consumption of cadmium contaminated food).   Other types of tissue
damage or functional disturbances have also been observed in man (e.g. bone,
haematopoietic system) and in animals (e.g. hypertension) and have also been
reviewed.   The available information suggests however that the critical organ
(i.e. the site of the initial lesion) is probably the kidney.   Evaluation of
no-effect levels and acceptable daily intake for the general population must
therefore be based on dose-response relationships for kidney damage.

2.1 Effect on the lung

Repeated or prolonged inhalation of cadmium dust or fume produced an ob-

structive pulmonary syndrome and possibly also emphysema.   The chronic
respiratory effects of cadmium were first observed by Friberg (127) and
since then have been reported by several authors investigating cadmium
exposed workers (3, 12, 16, 34, 44, 217, 219, 226, 253, 254, 297, 391,
514).

Removal from exposure does not necessarily prevent the evolution of the
disease (35, 129).   The concentration of airborne cadmium dust or fume
which prevents the development of this lesion is difficult to define for
several reasons.   Human information is usually derived from retrospective epidemiological surveys on workers and quantitative data concerning
their exposure is always incomplete.   Usually air analyses were made
only upon one or a few occasions and the assumption that no major changes
in the working conditions have occurred with time, is not necessarily
valid.

The particle size distribution of the airborne cadmium dust has rarely
been determined and it is known that the fraction of "respirable" dust is
generally lower in cadmium factories than in the general environment.
The use of respiratory masks further impairs the evaluation of the "true"
exposure of the workers.   In several studies only clinical examination
and chest X-rays were performed but no lung functional tests were reported.   The use of insensitive methods casts doubt about the validity
of the conclusions drawn from these studies (166, 268, 394, 504).

Cadmium fumes are considered more toxic than cadmium dust and the toxicity of the latter may depend on their chemical composition (which may
influence their solubility in lung tissue).   In men exposed for more
than 10 years to cadmium oxide fume in casting copper cadmium alloys,
Buxton (44) found increased residual volume.   At the time of his survey
the airborne cadmium concentration at the different work sites ranged
from 1 to $270\mu g/m^3$, the time weighted average exposure being certainly
below $100\mu g/m^3$ (226) but previous exposure was probably higher.   Examining 11 subjects exposed for a shorter period of time (7 to 11 years) to
cadmium oxide fume in the course of extracting cadmium from master alloys, Teculescu and Stanescu (485) found no lung function disturbances
compatible with emphysema.   The air concentration ranged between 1.21
and 2.70mg CdO/m^3.   Elevated excretions of cadmium in the urine were

noted, varying from about 3 to 65μg/24h. The short duration of exposure
may explain the negative findings.

Recently the ventilatory function of 144 workers exposed mainly to cadmium dust was evaluated and compared with that of a similar number of
control workers (253, 297). These workers belonged to four factories:
an electronic workshop, a nickel-cadmium storage battery factory and two
cadmium producing plants. The total airborne cadmium concentration and
the "respirable" fraction were determined at the different work sites
(sampling time usually 4 to 8 h.). Striking differences between total
and "respirable" dust concentrations were found. The ranges of total
dust concentration ($\mu g/m^3$) found in the four factories at the time of the
survey were: electronic workshop, 6.9 to 18.6; Nickel-cadmium battery
factory, 1 to 465; Cadmium producing plant A, 37 to 356; Cadmium producing plant B, 8 to 27,050. The highest respirable dust concentration
found in the same factories were: 4, 88, 21 and 65$\mu g/m^3$ respectively.

In the electronic workshop which occupies mainly female workers (average
duration of exposure: 4.4 years), no pulmonary disturbances were found.
Thus 4 years exposure to a time weighted average concentration of around
10$\mu g/m^3$ (total dust) or less than 4$\mu g/m^3$ (respirable dust) has probably
no deleterious effect on the lung. In male workers belonging to the
three other factories a statistical reduction in ventilatory capacity was
found in the group exposed for less than 20 years (average = 7.5 years)
as well as in the one exposed for more than 20 years (average: 27.5
years). The average total dust concentration to which these workers
have been exposed is highly speculative since as indicated above, the
concentration range is very large (mostly in the cadmium producing plant
B) and past exposure is not known precisely. However, since 88% of the
workers exposed for more than 20 years belong to the plant A it can be
estimated that a "respirable" dust concentration above 20$\mu g/m^3$ (8h/day –
5 days/week) can give rise to some degree of lung damage after several
years. This estimate is probably conservative since only workers still
in activity have been examined. This may have introduced a bias in the
sense that only "resistant" workers remained active.

It should, however, be stressed that the absolute change observed was
very slight and was only statistically significant because of the large

number of workers examined.   Indeed a recent extensive reevaluation of
the lung functions of 18 workers exposed to cadmium for more than 20
years and of 18 control workers failed to reveal significant differences
in the following lung function tests: lung volumes, 1-sec. forced expiratory volume, elastic recoil of the lungs, single breath lung diffusing
capacity, specific airway conductance (SGaW), maximal expiratory flow
rates (V max), slope of the N_2 alveolar plateau, N_2-closing volume and
closing capacity (Stanescu et al, to be published).

Unfortunately with the exception of the recent investigation carried out
by Baumert et al (26) animal experiments have only been performed with
relatively high exposure level (several mg/m^3) and do not shed light on
the no-effect level (127, 395).   Snider et al (456) reported changes re-
sembling human centrilobular emphysema in rats exposed to a 0.1% polydis-
perse aerosol of cadmium chloride in physiological saline (10 mg/m^3 - 1
hr each day for 5 to 15 times).

It has been found that tissues (kidney, lung, liver) of persons who die
from emphysema, chronic bronchitis and cancer of the lung contain larger
amounts of cadmium than subjects who did not have these diseases (180,
272, 316, 317).   As indicated previously, it is probable that cadmium
comes from cigarette smoke and these observations do not prove that cadmium was responsible for the lung lesions.

It is also possible that cadmium interferes with the lung defence mechanism since it has been shown that in vitro 5-10μM Cd^{2+} abolishes coupled
phosphorylations and respiratory control in macrophages (323).   At 0.1
mM Cd^{2+} reduces also rabbit alveolar macrophages viability in vitro
(519).   Recently Baumert et al (26) have evaluated the effect of cadmium
chloride aerosol inhalation on rat alveolar macrophages in vivo.   Groups
of 12 to 20 male rats were continuously exposed to 0.2 mg/m^3 $CdCl_2$
aerosol (aerodynamic diameter 0.32μm) for 7 or 66 days and to 0.05mg/m^3
(aerodynamic diameter 0.15μm) for 7 days.   At the end of the exposure
the animals were killed and the alveolar macrophages were isolated by
lavage.

In the group exposed to 0.2mg/m^3 for 66 days they made the following observations: 7 out of 20 rats died before the end of the experiment;

significant reduction (15%) in body weight, increase of lung weight; cadmium concentration in lung and kidney $\pm$ 40 ppm, cadmium concentration in liver: $\pm$ 9 ppm, emphysema associated with inflammatory changes in the lungs, increased number of alveolar macrophages, enhancement of their phagocytic activity and their number of lysosomes but reduction of their number of mitochondria which were swollen. In the group exposed to $0.2 mg/m^3$ for 7 days: increase of lung weight, cadmium concentration in lung: $\pm$ 11 ppm, kidney: 0.6 ppm, liver: 0.3 ppm, increased number of alveolar macrophages and of their phagocytic activity. In the group exposed to $0.05 mg/m^3$ for 7 days were found an accumulation of cadmium in lung of about 10 ppm (vs. 0.058 in control) kidney: 0.242 ppm (vs. 0.082) and liver: 0.099 ppm (vs. 0.031 ppm) and an increase of the phagocytic activity of the alveolar macrophages.

In summary

No long term animal experiments using sufficiently sensitive functional methods have been performed to define the airborne concentration of cadmium which has no deleterious action on the lungs.

The preliminary short term investigations of Baumert et al (26) are interesting; they should be completed by long term experiments. Epidemiological surveys on workers suggest that repeated 8hr exposure to concentrations above $20 \mu g/m^3$ "respirable" dust ($<5\mu$) could induce slight functional changes in susceptible individuals after less than 20 years exposure. Simple extrapolation would indicate that continuous exposure (24 hours/day for 70 years) to $2\mu g/m^3$ is probably close to the no-effect level for the lungs. The large degree of uncertainty of this estimate however must be stressed because it is based on very few data.

2.2 Effect on the kidney

In humans, cadmium induced kidney lesions have been observed in workers exposed to airborne cadmium and in Japan in persons (mainly old women) ingesting contaminated food (Itai-Itai disease) especially rice grown on soils irrigated with polluted water (239). It was again Friberg (126, 127) who first showed that prolonged exposure to cadmium dust causes kidney damage. Since then this toxic action of cadmium has been confirmed by many investigators who have surveyed workers exposed to cadmium dust or fume (3, 15, 34, 35, 219, 253, 297, 378, 379, 380, 381, 391, 411, 452, 453, 474, 505, 512).

In workers, kidney lesions usually precede lung damage (3, 253, 379). The classical kidney lesion involves the proximal tubule giving rise to a "tubular" type proteinuria (predominance of low molecular weight proteins like $\beta_2$microglobulin, lysozyme, ribonuclease, retinol binding protein and immunoglobulin chains). The presence of a low molecular weight albumin (minialbumin) has also been reported (223, 224). Sometimes a glomerular pattern (increased high molecular weight protein clearance) may accompany the tubular lesion (31, 253, 411).

It is also possible that proteinuria occurs or decreases after the workers have been removed from exposure (129, 345) but in general the proteinuria remains constant after removal from exposure (129, 383). The mechanism of the kidney lesion is still unknown. Chiappino and Pernis (62) have suggested that the intracellular catabolism of reabsorbed proteins in kidney tubules may be impaired before the tubular reabsorption of proteins. Cadmium proteinuria may be associated with:

- glucosuria (3, 35, 219, 380, 454, 474)
- aminoaciduria (3, 69, 150, 219, 380, 401).
  The increase in total excretion of aminoacids is considered as a relatively late and unspecific sign by some authors (31, 345) and an early biological manifestation by others (336). Increased excretion of citrulline, arginine and proline could occur earlier than changes in the total amino-nitrogen excretion (501, 518)
- decreased concentrating capacity of the kidney (4, 127)
- impaired acid excretion (219, 401)
- increased excretion of calcium and phosphorus (3, 219)
- change in the renal handling of uric acid (3, 219).

The majority of authors recognize that usually proteinuria precedes the other functional lesions (129). The disturbed metabolism of calcium and phosphorus may favourize the formation of renal stones which were prevalent in workers exposed to cadmium in Sweden and in the United Kingdom (3, 4, 129).

As indicated in Chapter V the renal cortex is an important site of cadmiu storage in the organism and therefore the concentration of cadmium in this organ increases with the intensity and the duration of exposure. However, when renal damage develops cadmium concentration in kidney decreases (probably due to increased urinary excretion). This phenomenon has been confirmed by animal experiments.

Since in man the kidney appears to constitute the critical organ during long term exposure to cadmium, it is of major importance to evaluate the no-effect level of cadmium on this organ. Several authors have suggested that the first detectable functional disturbance is increased low molecular weight protein excretion (e.g. $\beta_2$microglobulin). The no-effect level for this type of lesion should therefore be estimated (N.B. however increased high molecular weight protein clearance has been found to accompany or sometimes to precede the increased low molecular weight protein clearance. Up to now this observation in man has been reported by one laboratory only (31, 253, 411) and should be confirmed by further investigations. It is thus considered in this report that increased excretion of low molecular weight proteins represents the earliest biological manifestations of cadmium action on the kidney).

The animal investigations are first reviewed and then the results of various epidemiological surveys are considered.

A) Animal Data

Unfortunately in many animal experiments kidney function (in particular low molecular weight protein excretion) has not been investigated and it is known that histological changes may not necessarily be found when moderate proteinuria can already be demonstrated (127). Therefore this casts doubt on the conclusions drawn from animal experiments in which specific urinary proteins were not looked for. Nevertheless the main long term investigations carried out on animals can be briefly reviewed.

a. Oral administration

In rats receiving 31 ppm cadmium (as cadmium chloride) in their diet for 100 days no morphological changes were observed in tubules and glomeruli by Wilson et al (532). At 63 ppm slight morphological changes were found. Decker et al (82) administered cadmium (0.1 to 50 ppm) in drin-king water to male and female rats for 1 year. At 50 ppm reduction of blood haemoglobin was found. At 10 ppm they found no detectable adverse effects as judged by general health, growth rate, food and water intake, blood studies and tissue examination (kidneys) for pathological changes. Anwar et al (10) found no effect on the kidney in dogs receiving 0.5 to 2.5 ppm $CdCl_2$ in drinking water for 4 years and slight increase in renal and liver fat in dogs receiving 5 and 10 ppm $CdCl_2$. However the authors do not consider the morphological changes found in the kidney as due to cadmium. In rats receiving 7.5 ppm cadmium (as cadmium chloride) in drinking water and a calcium-deficient diet for one year, Piscator and Larsson (386) found an increase in ribonuclease excretion in urine (the mean cadmium concentration in renal cortex was then 90µg/g wet weight). In rats receiving 10 ppm cadmium in drinking water, electron microscopic examination of the kidney demonstrated slight changes already after 6 weeks and pronounced alteration after 40 weeks (330). 1 ppm of cadmium in drinking water for 24 weeks induces no electron microscopic changes in rat renal tubules (330). In rats receiving 10 ppm cadmium in food for 41 weeks, Sugawara and Sugawara (469) found slight tubular damages. The mean cadmium concentration in whole kidney was 91.4µg/g wet weight.

In those receiving 50 ppm only 12% survived for 41 weeks. The cadmium concentration in the kidney was then 217µg/g wet weight. Lorke and Loser (279) administered 0, 1, 3, 10 and 30 ppm cadmium (as $CdCl_2$) in the diet of rats for 3 months. They found no morphological and func-tional liver and kidney changes; low molecular weight proteins were not determined in urine. No hypertension developed in the rats. After 3 months the cadmium content of the kidneys of male rats (in µg/g probably wet weight) was:

0 ppm	:	0.13
1 ppm	:	0.49
3 ppm	:	1.06
10 ppm	:	4.98
30 ppm	:	12.06

Approximately the same results were obtained in female rats. The cad-mium concentration in the kidney is low suggesting a very low rate of cadmium absorption. Cadmium concentration in liver was still lower than in kidney. The results of Lorke and Loser (279) on cadmium concentration in rat kidney following oral administration of cadmium have been confirmed by Fowler et al (125) and Koller et al (241). The renal concentration of cadmium found by Fowler et al (125) in male rats receiving distilled drinking water containing from 0.2 to 200 ppm cadmium for 6 or 12 weeks is summarized below:

Mean renal cadmium concentration in male rats

Cadmium dose in drinking water (ppm)	Mean daily Cd consumption (μg/animal)	Cadmium concentration in kidney (μg/g)	
		after 6 weeks	after 12 weeks
0	–	1.13	0.83
0.2	5	0.64	0.28
2	50	2.40	2.53
20	500	9.03	11.38
200	2000	28.24	31.60

The same authors found also that kidneys of cadmium treated rats on a low calcium (0.1%) diet had higher content of cadmium than those on a normal calcium (0.7%) diet.

Lorke and Loser (279) repeated their experiments in dogs. Although clearance studies of PAH and inulin did not show any effect, urinary protein excretion in male dogs at 30 ppm and in female dogs in the 10 and 30 ppm groups tended to be higher than that found in the control animals of the same experiment. The authors have however indicated (personal communication) that the proteinuria found in all the treatment groups was within the normal range observed in Beagle dogs of their colony (female dogs: proteinura x ± 2 SD = 26.9 ± 29.8mg/100 ml). The cadmium concentrations in kidney and the protein concentrations in 24 hour urine found after 3 months are presented in the following table.

	Male dogs		Female dogs	
Dose (ppm)	Cd in kidney (μg/g)	Protein in urine (mg/100ml)	Cd in kidney (μg/g)	Protein in urine (mg/100ml)
0	0.60	25.1	0.38	20.9
1	1.30	24.7	1.45	30.1
3	2.60	32.7	3.40	34.3
10	12.40	25.1	7.19	46.0
30	15.45	36.7	17.15	47.0

Again cadmium concentrations in tissue are quite low and suggest a very low oral absorption rate of cadmium in dogs. The Institut für Toxokologie der Bayer AG has also undertaken a 2 year feeding study with cadmium in rats. Cadmium was administered at the dose of 1, 3, 10, 50 ppm. Currently only an interim report has been prepared; the results of the histopathological examination of the organs will not be available before the end of 1976. A slight growth reduction of the male rats and a temporary reduction of haemoglobin and haematocrit level in both sexes were observed in the 50 ppm group. All the other parameters measured at regular intervals during the study or at its completion (blood chemistry, proteinuria, glucosura, SGPT, SGOT and alkaline phosphatase in urine, electrophoresis of urinary protein, tissue weight, bone length, systolic blood pressure) were not different between the groups. On the basis of the available parameters the no-effect level was considered to be 10 ppm and at this level the kidney concentration was 15 ppm. This no-effect level is only provisional since the histopathological examinations of the organs are not yet available (Loser and Lorke, personal communication).

Nomiyama (336) administered cadmium orally to rabbits (300μg/g diet) and observed the first sign of kidney damage (enzymuria) when cadmium concentration in renal cortex reached 200μg/g. Ogawa et al (352) administered cadmium (146 ppm) in drinking water to mice for 90 days. They found that carbonic anhydrase activity was unaffected in blood, liver and kidney 10 days after administration but thereafter enzymic activity was reduced ($\pm$30%) in all three organs.

Recently Fowler et al (125) have reported that rats receiving 0.2 to 200 ppm cadmium in drinking water for 1.5 to 3 months developed constriction of small and median renal arteries. Cadmium concentration in the kidney of the 0.2 ppm group was less than 1µg/g wet weight (see above). However there was no evidence of dose-response relationships and the morphological changes were not more intense at 12 weeks than at 6 weeks. Changes in glomeruli and peritubular capillaries as well as a decrease in Bowman's space were observed. These effects were more marked in animals given the highest cadmium levels and appeared to be dose-related. By electron microscopy, glomeruli from kidneys of treated rats contained an apparent increase in mesangial matrix and glomerular capillary lumina were reduced in size. Some irregular thickenings of the glomerular capillary basal lumina were occasionally noted. There was also a slight but statistically significant dose-related increase in blood urea nitrogen at 6 weeks but not at 12 weeks.

In summary the morphologic findings of Fowler's study suggest that chronic cadmium administration produces constriction of smaller renal arteries and a diffuse fibrosis of capillaries at even low dose levels (0.2 ppm). These changes would suggest a decrease in effective renal circulation with time. Loser and Lorke have indicated that they could not reproduce Fowler's observation in Wister rats receiving 100 ppm Cd (as $CdCl_2$) in their diet for 3 months (personal communication). It should however be noticed that during his experiments on cadmium induced hypertension in rats (see below) Shroeder had also observed renal arterial and arteriolar lesions. According to Koller et al (241) 3 ppm cadmium in drinking water administered to mice for 70 days causes moderate to slight tubular and glomerular damage. The mean cadmium level in kidney was then 2.78µg/g wet weight. The results of the main animal investigations described above are summarized in Table 14.

T A B L E 1 4

CADMIUM NEPHROTOXICITY: SUMMARY OF ANIMAL EXPERIMENTS (ORAL ADMINISTRATION)

Dose (ppm)	Duration	Observations	Remarks	Reference
RATS				
0.1 (water)	52 weeks	-	Morphology	Decker 1958
0.2 (water)	6 weeks	+	Constriction of renal arteries	Fowler 1975
0.2 (water)	12 weeks	+	Constriction of renal arteries	Fowler 1975
1.0 (diet)	12 weeks	-	Morphology	Lorke 1974
1.0 (water)	24 weeks	-	Electron microscopy	Nishizumi 1972
2.0 (water)	6 weeks	+	Constriction of renal arteries	Fowler 1975
2.0 (water)	12 weeks	+	Constriction of renal arteries	Fowler 1975
3.0 (diet)	12 weeks	-	Morphology	Lorke 1974
7.5 (water)	52 weeks	+	Enzymuria	Piscator 1972
10.0 (water)	52 weeks	-	Morphology	Decker 1958
10.0 (water)	6 weeks	+	Electron microscopy	Nishizumi 1972
10.0 (diet)	41 weeks	+	Morphology	Sugawara 1974
10.0 (diet)	12 weeks	-	Morphology	Lorke 1974
20.0 (water)	6 weeks	+	Constriction of renal arteries	Fowler 1975
20.0 (water)	12 weeks	+	Constriction of renal arteries	Fowler 1975
30.0 (diet)	12 weeks	-	Morphology	Lorke 1974
31.0 (diet)	14 weeks	-	Morphology	Wilson 1941

T A B L E   1 4 (continued)

Dose (ppm)	Duration	Observations	Remarks	Reference
50.0 (water)	52 weeks	(Hb↓)	Hb↓ (morphology N$^-$[1])	Decker 1958
50.0 (water	6 weeks	+	Electron microscopy	Nishizumi 1972
50.0 (diet)	41 weeks	+	Morphology + mortality	Sugawara 1974
63.0 (diet)	14 weeks	+	Morphology	Wilson 1941
200.0 (water)	6 weeks	+	Constriction of renal arteries	Fowler 1975
200.0 (water)	12 weeks	+	Constriction of renal arteries	Fowler 1975
300.0 (water)	6 weeks	+	Electron microscopy	Nishizumi 1972
DOGS				
0.5 (water)	4 x 52 weeks	–	Morphology	Anwar 1961
1.0 (diet)	12	–	Morphology	Lorke 1974
2.5 (water)	4 x 52	–	Morphology	Anwar 1961
3.0 (diet)	12	–	Morphology	Lorke 1974
5.0 (water)	4 x 52	+(?)	Morphology	Anwar 1961
10.0 (water)	4 x 52	+−(?)	Morphology	Anwar 1961
10.0 (diet)	12	+−(?)	Morphology proteinuria (?)	Lorke 1974
30.0 (diet)	12	+−(?)	Morphology proteinuria (?)	Lorke 1974
MICE				
3.0 (water)	10	+	Morphology	Koller 1975

b.   Parenteral administration

-   in rabbits given subcutaneous injections of cadmium sulphate 0.65 mg/kg once a day 6 days a week, proteinuria was demonstrated after 2 months (127).

-   rats were given intraperitoneal injections of cadmium chloride (0.75mg Cd/kg thrice weekly) (36).   After 4 months histological examinations revealed kidney damage.   At that time renal concentration of cadmium was about 220-320µg/g wet weight (whole kidney).

-   rabbits were given subcutaneous injections of cadmium chloride (0.25mg Cd/kg 5 days a week).   After 11 weeks histological examination revealed mild degenerative changes in the proximal tubules.   The average cadmium concentration in the renal cortex was then 250µq/q wet weight (13).

-   in rabbits receiving subcutaneous injections of 2.25mg/Kg $CdCl_2$ 5 times a week for up to 7 months, Castano (49) found glomerular amyloïdosis after 2 months and tubular degenerative changes after 5 months.   The cadmium content of the kidney was not measured.

-   Nomiyama et al (334, 335) administered subcutaneous injections of cadmium chloride at dose levels of 15.6 and 1.5 mg Cd/kg day.   Loss of body weight and increase of urinary excretion of protein, alkaline phosphatase and acid phosphatase were found to be early warning signs suggestive of renal injuries.

c.   Inhalation studies

Friberg (127) exposed rabbits for 3 hours a day for 28 days per month for 8 months to 5 mg Cd/m^3 (as cadmium iron oxide dusts).   Proteinuria could be demonstrated after 4 months.   At the end of the experiment (8 months) no histological changes were found in the kidney.

B)   Human Data

Bonnel (34) examined 100 men who had been exposed to cadmium fume and found 16 cases with proteinuria among those exposed more than 5 years. At the time of the survey the average concentration of cadmium in the atmosphere of the workshop did not exceed 270μg/m^3 but previous exposure was probably higher (226).

Adams et al (3) examined workers from alkaline battery factories. The airborne dust concentration was usually below 500μg/m^3 with peaks up to 5 mg/m^3 in a department where workers were only present for a limited period each day. The analysis of cadmium oxide dust at this factory showed that only about 20% of the particles were in the respirable size range. Several workers had proteinuria and the latent period before the onset of proteinuria varies from 5 years to 24 years. The authors conclude that the threshold limit value for cadmium oxide dust, as for fume, should be 100μg/m^3 at most but stress the point that no data permits the conclusion that this level has no effect on men's renal function.

Tsuchiya (504) found that in workers exposed to cadmium fume (time weighted average concentration of approximately 0.130mg/m^3) for 9 months to 12 years proteinuria and anaemia could be detected. The author proposes a threshold limit value of 50μg/m^3 for cadmium fume.

Materne et al (297) found that in workers exposed to a "respirable" dust concentration probably above 20μg/m^3 (see section 2.1 for a discussion on cadmium concentrations at the workplaces) for 27 years, kidney lesion can be found in the majority of them. If one assumes 1) that 20μg/m^3 was the average time weighted "respirable" dust concentration, 2) that the workers inhale 10m^3 during 8 hour working day and worked 225 days a year, 3) that 16% of inhaled cadmium is absorbed (25% deposition, 64% absorption), 4) that the amount of cadmium excreted is negligible, it can be calculated that after 20 years their occupational activities would contribute 144 mg Cd to their body burden. Assuming that the contribution of other sources (food, smoking) to the body burden is approximately 30 mg, the total body burden would then be 174mg. If a third of this amount is concentrated in the kidney (average weight 300g) the theoretical concentration of cadmium in the kidney (in the absence of kidney damage) would be approximately 193.3μg/g wet weight (corresponding to about 290 ppm in the renal cortex). This value would be lower if excretion rate had been taken into consideration. For example if one assumes that the exposed workers excreted 10μg Cd per day the theoretical renal cortex concentration is then decreased to 168.3 ppm.

Kjellström et al (230) found that about 15% of workers with 9-12 years of exposure to about 50μg Cd dust/m^3 had tubular proteinuria as compared to 1% of those in the control group. This finding is in agreement with Materne et al (297) conclusion that exposure to 20μg Cd dust/m^3 for more than 20 years could still induce kidney lesion. Making the same assumption as above the theoretical renal cortex concentration of cadmium in these workers (10 year exposure) can be estimated at 350 ppm. If one assumes that they excreted about 10μg cadmium per day during the exposure period the theoretical cadmium concentration in the renal cortex is reduced to 290 ppm. These theoretical results are consistent (although they do not prove it) with Friberg et al's (129) estimate of the critical renal cortex concentration (200 ppm) (see below).

Surveying more than 100 workers exposed to cadmium, Lauwerys et al (253) observed that when urinary cadmium does not exceed 15μg/g creatinine, the probability of detecting tubular damage is low. Data obtained by Piscator and Nomiyama (cited in ref. 129) agree quite well with this observation. Glomerular type proteinuria could, however, be found in workers with cadmium in urine as low as 7μg/g creatinine. It should, however, be stressed that in workers high urinary cadmium concentration can be found without proteinuria. In an examination of death certificates of a rural district with two copper workers in U.K., Davies (79) found that 3 times the expected deaths from nephritis and nephrosis occurred between 1948 and 1970. The author indicates that the excess deaths from renal disease probably was due to unrecognized renal cadmium poisoning.

Several studies performed in Japan suggest that there is a close correlation between cadmium concentrations in soil and Itai-Itai disease (renal tubular damage accompanied by osteomalacia and osteoporosis found mainly in old women who had borne several children) and prevalence of proteinuria and glucosuria (133, 215, 540). In women over 40 years the "normal" prevalence of proteinuria (sulphosalicylic acid method) ranges between 5 and 25% depending on the age (with an average of ± 15% at 55) (132). This prevalence reaches ± 20% when cadmium in soil is ± 2μg/g (133).

In the Jintsu Valley of Japan where Itai-Itai disease was prevalent, Yamagata and Sigematsu (540) estimated that the cadmium concentration in rice ranged from 0.72 to 3.87 ppm whereas in control areas the concentration ranged from 0.07 to 0.11 ppm. They concluded that the daily intake of cadmium by the Itai-Itai patients was probably about 600µg or above (assuming an average cadmium concentration in rice of 1 ppm).

After 20 years such a diet could thus produce a theoretical concentration of cadmium in renal cortex of approximately 438 ppm (neglecting excretion and contribution from other sources and assuming an oral absorption rate of 6%). It should be pointed out that these patients suffered also from nutritional deficiencies which may have exacerbated cadmium toxicity.

Friberg et al (129) have also extensively reviewed the results of various epidemiological surveys carried out in Japan. They concluded that it was impossible to use the data to estimate a no-effect level. I have, however tried to tabulate the average oral cadmium intake and the range of cadmium in urine found in the polluted areas where an increased preva- lence of proteinuria seems to have been discovered and in the correspon- ding control areas (Table 15). These results suggest that an average oral daily intake of approximately 200µg Cd is probably close to the no- response level for proteinuria. Such an intake would apparently be as- sociated with a urinary cadmium excretion of 6µg/L (Table 15). A daily intake of 200µg would correspond after 50 years to a theoretical body burden of 219 mg (assuming no excretion and an absorption rate of 6%) or a theoretical renal cortex concentration of 364 ppm. This estimate would be lower if excretion had been taken into consideration.

A recent report prepared by Shigematsu (444) dealing with the correlation between the prevalence of proteinuria and glucosuria and the degree of rice pollution in Japan suggests that the prevalence of one or both bio- logical signs might be increased in hamlets where the average concentra- tion of cadmium in rice is 0.7 ppm. In the hamlets where the cadmium concentration in rice is on the average 0.338 ppm, the effect is negli- gible or at the borderline.

Following Friberg's assumption (129) that on the average half the daily cadmium intake comes via rice and the daily intake of rice is 300 g, the daily average intake of cadmium would be 420µg in the hamlets where cad- mium in rice reaches 0.7 ppm and 203µg in hamlets where cadmium in rice is approximately 0.338 ppm. The last estimate of the daily intake (200

T A B L E   1 5

Epidemiological Surveys in Japan

| | Polluted areas probably with high prevalence of proteinuria | | Non polluted areas | |
	Oral cadmium intake µg/day	Cadmium in urine µg/L	Oral cadmium intake µg/day	Cadmium in urine µg/L
1. MIYAGI	245	7.1 – 10.59	85	4.64
2. TSUCHIMA	213 – 215	10 – 30	53 – 104	2 – 6 (?)
3. KAKEHASHI	480	10 – 23		
4. IKUNO	290 – 410	7.5 – 13.8* 6 – 14 * 11.9 – 35 *		
5. FUCHU	600	10 – 40	60	

* Persons with proteinuria

Adapted from reference 129.

μg) which corresponds to the borderline effect on the kidney agrees quite well with that based on previous epidemiological surveys carried out in Japan.

Recently Kjellström et al (231) attempted to elaborate a dose-response relationship between oral cadmium intake and cadmium induced tubular proteinuria among Japanese women aged 50-59 years living in cadmium polluted areas. Dose was estimated as the product of household rice cadmium concentration multiplied with resident time at present address plus a background dose during resident time outside polluted areas. Measurement of $\beta_2$microglobulin in urine was selected as the biological parameter for tubular proteinuria. The authors concluded that an average dose between 6 and 12 (years x μg/g) in a population would give an increased prevalence of high $\beta_2$microglobulin excretion as compared to control groups. This dose corresponds to 30 year-exposure to 0.2-0.4μg cadmium/g rice or a daily oral intake of 120-240μg cadmium (assuming a daily intake of 300g rice which corresponds to 50% of total oral intake of cadmium).

It should be stressed that factors other than cadmium exposure may have played a favourizing role in the development of Itai-Itai disease. Recently Nomiyama et al (337) have suggested that although long-term exposure to low concentrations of environmental cadmium coupled with deficiencies of calcium and protein in the daily diet has played a role in the etiology of Itai-Itai disease, the renal dysfunction found currently in Itai-Itai disease patients might have resulted from administration of large doses of vitamin D. However, Nogawa et al (333) concluded on the basis of their recent observations in the Hyogo Prefecture that cadmium is the main etiologic agent of Itai-Itai disease.

Comparing the renal cortex concentration of persons excessively exposed to cadmium and who did or did not present signs of kidney damage (± 30 cases), Friberg et al (129) concluded that 200μg Cd/g wet weight in renal cortex may be the critical concentration at which level tubular dysfunction may appear in a certain percentage of the population. Friberg et al (129) have also proposed a model to calculate the cadmium intake by different routes which is necessary to reach this critical concentration in kidney cortex. The model is based on the following assumptions:
-    one third of the body burden is in the kidneys (concentration in

cortex 50% higher than the renal mean value).

- of the absorbed cadmium 10% is considered to be rapidly excreted and this fraction is excluded from the calculation because it ultimately could contribute less than 1% of the total body burden of cadmium.
- 4.5% of ingested cadmium and 25% of inhaled cadmium are absorbed.

The model can take into consideration variation of kidney weight and calorie intake with age. Friberg and his collaborators (129) estimated that the excretion rate is in the order of 0.005% or less. This value gives the most plausible intake figure to reach a renal cortex concentration of 50µg/g wet weight at at 50 (value which has been found by analysis at autopsy samples). Considering first the oral intake and neglecting intake via ambient air or smoking, Friberg (129) calculated that the necessary daily cadmium intake to reach the critical concentration of 200 ppm in the kidney cortex at age 50 is 248µg. It can also be estimated that the adult oral intake of cadmium necessary to reach 50 ppm in renal cortex at age 50 is 62µg/day. Considering only the respiratory intake, the necessary cadmium concentration in ambient air to reach the critical cadmium concentration in kidney cortex at age 50 is 2µg/m^3 (ventilation 20m^3/24 h). The Subcommittee on the Toxicology of Metals under the Permanent Commission and International Association of Occupational Health has recently endorsed Friberg's proposal of 200µg cadmium/g wet weight as the tentative critical concentration in human kidney cortex (344). However, the tentative character of this proposal should be stressed because it is based on few observations on humans and recent experimental results (Fowler et al, Koller) suggest that kidney impairment can occur at a lower kidney concentration.

In summary on the basis of the available information, the best estimate of the no-effect level of ingested cadmium on the human kidney is 200µg cadmium daily (assuming that the absorption by the pulmonary route is negligible). Sufficient data on groups ingesting between 100 and 200µg daily are still too limited to conclusively confirm the validity of this proposal. With regard to the critical level of cadmium in human kidney the value proposed by Friberg et al (129) 200µg/g wet weight renal cortex has been tentatively accepted by several workers. Some animal data (derived mostly from experiments in which cadmium was administered parenterally) tend also to confirm this proposal. Unfortunately the results of other experimental investigations in which cadmium was administered

orally are more controversial. Some recent reports suggest that kidney
concentrations even below 3µg/g wet weight are associated with morphological changes in the kidney of rats and mice. To the contrary, other
authors did not evidence any change in rat and dog kidney when the concentrations exceeded 10µg/g.

2.3 Effect on the cardiovascular system (in particular hypertensive effect)

Schroeder and his collaborators have reported that it is possible to induce hypertension in rats by long term administration of cadmium (5 ppm)
in deionized water to which essential elements were added (210, 425, 427,
431, 435). The latent period before hypertension becomes manifest can
last up to one year. The cadmium concentration in kidney and liver of
hypertensive rats was about the same as in American adults (i.e. $\pm$40µg/g
and 6µg/g respectively). Schroeder and co-workers (437) have
also claimed that even the small amount of cadmium present in the commercial diet fed to growing rats for three months was enough to raise their
blood pressure. Statistically significant increase in blood pressure
could still be measured in rats fed diet containing 0.56-0.63 ppm cadmium
by comparison with animals fed special low cadmium diets containing 0.02
ppm cadmium (210). Repeated intraperitoneal administration of cadmium
(2 mg/kg cadmium acetate once a week for 7 weeks or over 9 months) can
also cause hypertension in rabbits (489, 492).

Thus Fischer and Thind (117) have produced hypertension in rabbits by injection of cadmium acetate 2 mg/kg intraperitoneally at weekly intervals.
The animals were killed when the elevation of the ear blood pressure to
above 200 mm Hg was maintained for a period of one or two weeks. Seven
of the eight rabbits received an average of 15.77 mg cadmium acetate and
the eighth rabbit received a total of 39.3 mg of cadmium acetate. The
cadmium concentration found in the tissues of control and hypertensive
rabbits is summarized in Table 16.

T A B L E   1 6

Tissue concentration (mean in μg/g tissue)

	Normal	Hypertensive
Kidney	0.290	35.641
Liver	0.077	33.360
Aorta	0.063	0.931
Mesenteric Artery	0.075	1.861
Pulmonary Artery	0.139	1.188
Heart	0.021	0.641

(from reference 117: see text for explanation)

The authors conclude "that the presence of cadmium ion in the blood vessels at concentrations of about 1.0μg/g wet tissue and in the kidney at a
much higher concentration may have a role in the pathophysiology of cadmium hypertension".   After discontinuation of cadmium exposure blood
pressure returns progressively to normal (433).   In rats cadmium can
also exacerbate renal ischaemic hypertension (321).

Recently Perry and Erlanger (374) have reported that cadmium could have a
biphasic action: at low doses (1-5 ppm in drinking water) it exhibits a
pressor action while at higher doses (50 ppm in drinking water) it decreases systolic blood pressure.   Thind et al (493) were also able to
induce a slight degree of hypertension in the dog by intraperitoneal injection of cadmium acetate (2 mg/kg I.P.).   Throughout the study period
of 42 weeks there were no obvious clinical symptoms and signs of cadmium
overdosage.   The only evidence of renal function impairment was a decrease in the creatinine clearance.   In all tissues analysed at the end
of the experiment (kidney cortex, kidney medulla, liver, ascending aorta,
descending aorta, thoracic aorta, abdominal aorta, coronary arteries,
pulmonary artery, carotid artery, main renal artery, proximal mesenteric
artery, distal mesenteric artery, mesenteric artery branches, spleen,
heart, diaphragm), cadmium concentration was significantly increased but
cadmium sequestration was most marked in the kidney.   A slight increase
in the hepatic and renal concentration of zinc was also found, a finding
that the authors (494) interpret as a protective reaction by the body in
an attempt to counteract the very large deposition of cadmium in these

organs.   The cadmium concentration in the kidney cortex of hypertensive
dogs was about 800μg/g dry weight which corresponds to approximately
230μg/g wet weight.

Schroeder's group has suggested that the cadmium to zinc radio in kidney
plays an important role in the development of hypertension.   In rats a
ratio above 0.8 was always associated with hypertension.   The value of
this ratio could therefore be more important than the concentration of
cadmium alone.   The administration of a chelating agent (cyclohexane -
1,2,-diamine NNN'N' - tetraacetic acid) which can bind cadmium more
easily than zinc can reverse the cadmium induced hypertension in rats.
It is believed that the lowering of the cadmium zinc ratio in kidney is
responsible for this effect (432, 433).

Several investigators have also shown that cadmium may enhance sodium reabsorption in renal tubules either after single administration (54) or
after repeated exposures (264, 372).   Lener and Bibr (263, 266) could
not repeat Schroeder's observation with regard to the hypertensive effect
of cadmium in rats but they (266) could induce hypertension in rats if
the animals received 2% NaCl in drinking water before cadmium treatment.
Porter (389) was also unable to produce hypertension in rats with cadmium
under a variety of treatment schedules and conditions.   Frickenhaus et
al (130) could not find significant changes in blood pressure in female
rats given cadmium sulphide in their diet for 3 months (26 and 52 ppm).
Strain difference in susceptibility to cadmium may be responsible for
these discrepancies.

The basic mechanism by which cadmium can induce hypertension in animals
is still highly speculative.   Disturbance of the sodium balance by
increased tubular reabsorption (509) possibly by increased plasma renin
activity and aldosterone secretion has been suggested (369, 372).
Interference with vasopressin action has also been hypothesized.   Bentley et al (28) have observed that cadmium (10^{-5} M) reduces the hydro
osmotic response of the toad urinary bladder to vasopressin in vitro but
has no effect on the natriferic response of this hormone.   Cadmium did
not change the osmotic permeability of the toad bladder in the absence of
vasopressin suggesting that Cd^{2+} may primarily effect the vasopressin-
induced increase in osmotic water movement.   Cadmium, however, also

inhibited the hydro-osmotic effect of added cyclic AMP so that it seems likely that the metal ion is acting at a stage of the stimulatory mechanism after the endogenous formation of the nucleotide.

In a study of the vascular reactivity of the aortic strips obtained from normal rabbits, Thind et al (489) found that cadmium produced a dose-related reversible inhibition or reversal of the angiotensin epinephrine and levarterenol (norepinephrine) responses. They also showed that the vascular responsiveness to angiotensin but not to levarterenol was significantly decreased in the subacute phase of cadmium hypertension and that the hypertensive rabbit aorta strips exhibited decreased stiffness as compared with those from normal rabbits (490). The same authors found that injection of cadmium acetate into the renal artery preceding angiotensin, epinephrine and levarterenol resulted in a dose-related reversible inhibition of the vasopressor-induced renal vasoconstriction in the absence of systemic haemodynamic changes in the dog (488, 491).

Fischer and Thind (117) have demonstrated that cadmium can accumulate in vascular tissues of rabbit treated with cadmium (see above) and have suggested that the cadmium effect in the vascular tissue could be mediated through changes in the contractile mechanism at the vascular level. Furthermore, the reduced vascular responsiveness to angiotensin resulting from such action could produce an alteration in the renin-angiotensin feedback system in the kidney where cadmium deposition is the greatest (117).

Perry and Erlanger (373) have also found elevated circulating renin activity in rats following doses of cadmium known to induce hypertension. Recently Amacher and Ewing (6) have found that in dogs receiving 3 intravenous injections of cadmium chloride (1 mg Cd/kg body weight per injection) the renal artery has the greatest affinity for cadmium of the ten major arteries examined. They suggest that if the predilection of the renal arteries for cadmium deposition extends to renal arterioles as well, the intrarenal vascular tissue may play a more sensitive role in cadmium hypertension than the peripheral vascular tissue.

We have already discussed in the preceding section (22) the observation of Fowler et al (125) who reported that rats receiving 0.2-200 ppm cadmium in drinking water for 1.5-3 months developed constriction of small and median renal arteries. On the basis of his observations on animals and the results of a clinical study indicating that patients with hypertension excreted more cadmium than did controls (367), Schroeder has suggested that cadmium contributes to the high prevalence of hypertension in the U.S. (425, 427, 430, 431) but other researchers question the existence of a causal relationship between cadmium and hypertension in man. Schroeder (427) reported that in several countries persons dying from hypertension had higher kidney concentration of cadmium and higher cadmium to zinc ratio than those dying of other causes.

Thind (492) reported that the mean plasma concentration of cadmium in 32 hypertensive persons was significantly higher than in 15 normal subjects. McKenzie and Kay (301) found a higher urinary excretion of cadmium in hypertensive women than in normotensive control. In Czechoslovakia, Lener and Bibr (265) have found that cadmium concentration in kidney of 12 men who died from hypertension was 36.11μg/g while in a control group the average concentration was 27.02μg/g.

Voors et al (513) reported that aorta cadmium levels and kidney cadmium - zinc ratios were somewhat higher in patients with cardiovascular disease (atherosclerosis and hypertension) (CVD) as compared to non-CVD patients. Karlicek and Topolcan (214) found higher cadmium concentrations in 20 patients who died from essential hypertension (by comparison with 20 victims of accidents). Glauser et al (147) found that 10 living normal humans (mean age: 30.9 years) had a blood-cadmium level of 0.3μg/100ml while a matched group of 17 living untreated hypertensive humans had a blood cadmium of 1.1μg/100 ml. The difference was statistically significant. All of the normal subjects had blood cadmium levels below 0.8μg/100 ml while 13 of the 17 hypertensive patients had blood cadmium levels over 0.8μg/100 ml.

Bierenbaum et al (33) have reported that blood pressure and serum cadmium were significantly higher in Kansas City, Kansas than in Kansas City, Missouri and suggested that both phenomena were related. In view of the preferential accumulation of Cd in red blood cells, their mean values for serum cadmium appear very high (0.12μg per 100 ml serum in Missouri vs. 1.64μg per 100 ml in Kansas). This raises doubt about the validity of their analyses.   The statistical handling of their data has also been criticized.   Examining the urine of 27 subjects, Mertz et al (307) found a tendency for an increased excretion of cadmium in hypertensive subjects but their results on urinary cadmium concentration appear too high by a factor of 10 or more.

Carrol (47) found a correlation between cadmium concentrations in the air of 28 American cities and death rates from hypertension and arterioscle-rotic heart disease.   Hickey et al (177) have also found a good correla-tion between atmospheric pollution by cadmium and death from cardiovascu-lar diseases in 26 American cities.   The correlation could be signifi-cantly increased if vanadium was considered together with cadmium.   It should, however, be recalled (see Chapter V) that usually air constitutes a minor source of cadmium by comparison with food.

Since an inverse correlation between the prevalence of cardiovascular diseases and the hardness of drinking water has sometimes been found (75, 293, 319, 428) and since hard water usually contains less heavy metals, in particular cadmium, than soft water does, one may also find a correla-tion between cadmium content of water and prevalence of hypertension (430, 432).

It should however be stressed that potential association of hypertension with water hardness is still controversial (5, 275, 320) and that, like for air, the intake of cadmium with water is rather low.

Masironi (294) has stressed the fact that various human populations show statistically significant differences in renal cadmium concentration, with Caucasoid Americans, Europeans and Asiatics having more than Negroid Africans and less than Mongoloid Asiatics.   This pattern is consistent with the epidemiology of hypertension, namely that both renal cadmium and blood pressure are low in Africa, south of the Sahara and high in parts of eastern Asia particularly in Japan and Formosa.

It has also been speculated that the increase of deaths from ischaemic heart disease observed in recent years in Scandinavian countries (mainly Norway) could be brought about by increased industrialization and mainly higher pollution with Cd which, being accumulated by shellfish and other

marine animals is eventually ingested by humans (292). An increase of
the cadmium level in fish and shellfish in Norway has not been reported,
however. Furthermore, several researchers have found no relationship
between cadmium and hypertension in man.

In the USA, Morgan (318) could not find a higher renal concentration of
cadmium in 12 negroes who died from hypertension than in 20 control
subjects. Hine et al (179) and McKenzie (302) have found no correlation
between cadmium concentration in human tissues and hypertension. Hunt
et al (190) studying 77 cities in the US could not find a significant
correlation between cadmium fallout and mortality from cardiovascular
disease and furthermore by reanalysing Carroll's data they show a higher
correlation between population density and hypertension than between
hypertension and cadmium concentration in air. Finally it should be
mentioned that hypertension is not a common feature among Itai-Itai
patients (332) nor in workers exposed to cadmium (34, 35, 127, 186, 219,
297). This negative finding must perhaps be linked with the Perry and
Erlanger (374) suggestion mainly that at low doses cadmium exhibits a
pressor action while at higher doses it decreases systolic blood pressure.

2.4 Effect on the bones

Osteomalacia and osteoporosis with a tendency to fracture and bone deformation accompanied by lumbar pains, leg myalgia and pains on bone pressure as well as disturbance of gait have been described in Itai-Itai
patients principally in women after menopause who had borne several
children (443). It is believed that nutritional factors (in particular
calcium deficiency) have exacerbated the disease. The levels of calcium
and inorganic phosphorus in serum are often low while the alkaline phosphatase level is high (131). The same type of bone lesion has also been
found in workers exposed to cadmium (in particular pseudofracture) (3,
141, 328, 401). In workers, low (Vorbieva, personal communication) or
high (219) calciuria has been reported.

Bone lesions can also be induced in animals with cadmium. Itokawa et al
(199) induced osteomalacia in rats by giving them 50 ppm cadmium in water
along with a calcium deficient diet. An abnormal curvature of the

spinal column was found in rats fed a low protein, calcium deficient diet containing 200 ppm cadmium (198).   Kawai et al (216) could also produce decalcification and cortical atrophy of bones in rats fed 10 ppm cadmium in their diet.   Since cadmium does not particularly accumulate in bones (30, 343), it has been first suggested that the bone lesion is not the result of a direct action of cadmium but is probably due to a disturbed calcium and phosphorus metabolism secondary to the kidney lesions (3, 129) possibly associated with a disturbance in vitamin D metabolism.

It has been shown that 1. hydroxylation reaction of 25-OH-vit.D_3 by chick kidney mitochondria is inhibited by 0.1 mM cadmium in vitro.   This inhibition is prevented when cadmium is bound to metallothionein (468). However Kimura et al (225) have reported that the in vivo 1. hydroxylation of 25-hydroxycholecalciferol proceeded without appreciable inhibition even in rats loaded with large amounts of oral cadmium.   Furthermore, Bowden and Hammarström (27) have found that in young animals, injected cadmium can accumulate in the osteoblasts thus raising the possibility of a direct toxic action of cadmium on bone tissue (see also other lesions).   This suggestion has been reinforced by the observations made on young rats fed a diet containing 10 to 300 ppm Cd (as cadmium chloride) for 3 weeks and which developed bone lesions (osteoporosis) before the occurrence of histological changes in the kidneys.   It should however be stressed that this investigation did not evaluate the functional status of the kidney (545).

Yushas et al (546) reported also that cadmium (100 ppm) administered to rats in their drinking water for up to 6 weeks can alter calcium deposition in bone without concurrent change in urinary calcium excretion.   In animals, a low calcium diet increases the absorption of cadmium (251) and cadmium (50 ppm in drinking water) can depress calcium absorption from the intestine.

Some Japanese authors have stated that Itai-Itai disease is not caused by cadmium but is mainly a nutritional osteomalacia caused by the insufficient intake of vitamin D, as well as other malnutritional factors (208).

2.5 Effect on the haematopoietic system

Slight hypochromic anaemia has been seen among most Itai-Itai patients as well as among workers exposed to cadmium (127, 297, 328, 383, 504). Anaemia has also been produced in animals by several investigators (82, 118, 127, 398, 554). In rabbits, anaemia may be associated with eosinophilia (127). Simultaneous administration of iron can prevent the development of anaemia (123, 124, 128, 387). In quail, vitamin C prevents the anaemia induced by dietary cadmium (75 ppm for 4 weeks) (123). According to Decker et al (82), 10 ppm cadmium in drinking water for 1 year does not cause anaemia in rats while Wilson et al (532) found that 31 ppm for a couple of months is already effective.

It has been demonstrated that short term (8 days) administration of cadmium (10 ppm or higher doses) to mice inhibit partially the gastrointestinal absorption of iron. Itokawa et al (199) found also fatty deposition in bone marrow of cadmium treated rats (50 ppm in drinking water for 120 days).

2.6 Effect on the liver

Effects on the liver have been described in animals repeatedly exposed to cadmium. Cirrhotic changes were found by Friberg (127) in rabbits injected with cadmium sulphate (0.65 mg/kg 6 days a week for 2 to 4 months) Colucci et al (70) injected rats with $CdCl_2$ (route of injection not specified, 0.5 to 4 mg/kg daily for 6 or 7 days). They found that when the hepatic cadmium level was below 30µg Cd/g wet weight (corresponding to a dose below 2µg/kg) the metal was sequestered as metallothionein and no evidence of morphological liver damage was apparent.

Stowe et al (466) discovered morphological changes in liver of rabbits receiving 160 ppm cadmium in drinking water. Liver function tests (SGOT, SGPT, LDH, BSP retention, blood coagulation tests) were not altered. After 6 months administration of cadmium its liver concentration reached 188µg/g wet weight. Sporn et al (459) found that in rats 1 ppm cadmium in drinking water for 1 year induced a change in the activities of some liver enzymes. Miller et al (310) found morphological changes in liver of rats receiving 17.2 ppm cadmium in drinking water. Intra-

peritoneal injections of $CdCl_2$ (1 mg/kg) for 45 days to rats stimulate
the gluconeogenic pathway in liver (and in kidney) (450). The same
treatment impairs respiratory control in isolated liver mitochondria
(89). Slight changes in liver tests (Takata, thymol, increased serum
gammaglobulin) have been reported in a few workers exposed for 20 years
to cadmium oxide dust (127).

2.7 Effect on animal growth and survival

In several animal experiments the effect of long term cadmium administration on growth and/or survival was also evaluated. Fitzhugh and Meiller
(118) found no effect on growth rate in rats on a diet containing 15 ppm
cadmium. Schroeder, Vinton and Balassa (423) found that the growth
curves of male mice receiving 5 ppm oral cadmium did not differ in any
obvious way from those of the "cadmium free" control mice. There was,
however, a significantly increased mortality rate of the male mice at 6
months, 21 months and 24 months of the oral cadmium feeding. Thind et
al (489) made the same observation in rabbits treated with cadmium (no
change in body weight but slight increase in mortality).

Recently Exon et al (105) have suggested that cadmium and the protozoam,
Hexamita muris, could act synergistically in causing the death of mice.
They found that 4- to 5-week old mice exposed to 300 or 3 ppm cadmium as
cadmium chloride in the drinking water suffered 26 and 7% mortality respectively. Death did not occur in control mice reared in the same room.
Clinical signs and histopathology established Hexamita muris as the
causative agent. They therefore suggest that cadmium may have decreased
the natural defence mechanisms of the animals. Decker et al (82) found
that concentrations of cadmium ranging from 0.1-10 ppm in aqueous solution did not significantly alter growth patterns of rats. At 50 ppm
reduction of growth rate was observed. Significant stunting was found
when cadmium was fed as a dietary supplement in concentrations of 31 ppm
(532). The same effect was noted when dietary concentrations were
increased to 100 ppm (19). Pribble and Weswig (392) reported that the
growth rate of rats receiving 5 ppm cadmium in water was reduced but this
effect was not observed when the 5 ppm cadmium was administered in the
diet. Accumulation rates of cadmium in the liver and the kidney were

also higher in rats receiving cadmium as an aqueous supplement than in
those receiving the same level as a solid supplement.

2.8 Other effects

Anosmia has been reported in a group of alkaline battery workmen who were
exposed to both cadmium and nickel dust (2, 15, 391).

Yellow coloration of teeth has been observed among workers exposed to
cadmium dust or fume (394, 552).   Since inhibition of pigmentation of
the rat incisor has been identified as a sensitive indicator of cadmium
toxicity in that species (532) and since teeth in rats are rendered less
resistant to caries as a result of cadmium supplementation (144, 260),
Bowden and Hammarström (27) have studied the distribution of labelled
cadmium injected intraperitoneally into 10-day old rats with particular
attention directed to the dental tissues.   They found that the injected
cadmium was accumulated in the cells forming enamel, dentine and bone
(osteoblasts) which suggests that the toxic effects of cadmium on these
tissues are due to a local cellular change in addition to the systemic
effects on calcium and iron metabolism which have also been described in
the literature.

According to Murata et al (cited in Ref. 142), a decrease in pancreatic
function has been found in Itai-Itai patients.   Impaired pancreatic
function and morphology (20, 142) and increased adrenal catecholamine
levels (404), have also been found in animals receiving repeated administrations of high doses of cadmium.

Recently Koller et al (242) have found that 3 ppm cadmium administered in
drinking water for 70 days can induce immunosuppression in mice and this
effect persists for several weeks after discontinuance of exposure.   The
same phenomenon was observed in rabbits receiving 300 ppm cadmium chloride in their drinking water for 70 days (240).   Reduction in the activity of intestinal brush border, alkaline phosphatase and ATPase was
found in rats receiving 100 ppm cadmium (as $CdCl_2$) in their drinking
water for 30 days (470).

Chowdhury and Louria (66) reported that in vitro Cd at a concentration above 10µg/ml decreases the α_1 antitrypsin content of plasma and they speculated that this action could be responsible for the emphysema reported in some industrial workers exposed to cadmium. However, the concentration active in vitro is much higher than that found in exposed workers. Furthermore, (Bernard et al, unpublished results) recently measured α_1 antitrypsin in serum of 20 workers exposed for more than 20 years to cadmium and could not find reduction in their serum α_1 antitrypsin concentration.

VI - 3   CARCINOGENICITY

Cadmium and its compounds has been conclusively shown to induce sarcomas at injection sites in animals. By intramuscular injection of suspended cadmium metal powder Heath and his collaborators (173) were able to induce rhabdomyosarcomas at injection sites in rats. Zinc and tungsten metal powders injected in the same manner produced no tumours.

Gunn et al (153) found pleomorphic sarcomas at injection sites and interstitial cell tumours in rat testis, following a single injection of cadmium chloride (0.03 mmole/kg). The tumouregenesis action of cadmium could be prevented by 3 subcutaneous administrations of zinc acetate (1 mmole/kg). Single subcutaneous or intramuscular injection of cadmium chloride in amounts as low as 0.17 to 0.34 mg of cadmium induced sarcoma (154).

Roe and his collaborators (160, 410) found that of 20 rats given ten once-weekly subcutaneous injections of 0.5 mg cadmium sulphate in water, 14 developed sarcomas at the site of injection within 20 months after the start of treatment. All showed testicular atrophy and many showed interstitial-cell hyperplasia. Of 18 rats examined post-mortem within 20 months of the start of treatment 10 had interstitial-cell tumours. Knorre (235, 236, 237), Lucis et al (285) and Reddi et al (407) also observed sarcomas at the injection site and interstitial-cell tumours in rats which had received a single subcutaneous injection of cadmium chloride (0.03 mole/kg). Other cadmium compounds (cadmium sulphide, cadmium oxide, cadmium-precipitated rat ferritin) can also cause sarcomas at injection sites (160, 218, 220, 410). It should, however, be stressed that many substances when similarly administered to rats can induce tumours at injection sites.

Malcolm (290) reported results from a two year mortality experience of rats given cadmium sulphate subcutaneously (0.2 to 0.05 mg at weekly intervals) and orally (0.8 to 0.2 mg at weekly intervals) via stomach tubes, and mice given cadmium sulphate orally (0.02 to 0.005 mg/kg body weight at weekly intervals). Sarcomas were observed in a few rats given subcutaneous injection but the incidence of testicular tumours was not different between control and exposed animals.

Repeated once weekly subcutaneous injections of 0.05, 0.1 or 0.2 mg/rat cadmium sulphate or repeated gastric instillation of 0.2, 0.4 and 0.8 mg/kg cadmium sulphate to rats for two years failed to produce neoplastic change in the prostate gland (269, 270). Cadmium sulphate administered orally to mice (up to 4 mg/kg each week for eighteen months) failed to produce neoplastic or preneoplastic changes in the genito-urinary tract generally or in the prostate gland in particular (271). Publications by Schroeder and his collaborators suggest also that oral administration of cadmium (5 ppm cadmium acetate in drinking water) has no carcinogenic action (209, 424, 426). It has, however, been stressed that the experimental level used, which was designed to simulate the human exposure level, was too low for carcinogenic valuation (192).

The first report of a potential cancer risk of cadmium to man is due to Potts in 1965 (391). He reported eight deaths from a group of 74 workers exposed for over 10 years to cadmium oxide dust. Three of the deaths were due to carcinoma of the prostate, one was due to carcinoma of the bronchus, and one was due to carcinomatosis. However, due to the small number involved, Potts did not conclude any relationship.

In 1967, Kipling and Waterhouse (227) investigated 248 cadmium workers who had been exposed for at least one year to cadmium oxide. They found a high incidence of prostatic carcinoma (4 cases vs. an expected number of 0.58) but the total cases of malignancies (12) were close to expected figures. In 1968, Humperdinck (189) reported 5 carcinomas from a group of 536 relatively young workers who, from 1949 to 1966, had any contact with cadmium. The author did not find any causal relationship between cadmium exposure and cancer. It should be stressed that for the majority of workers, duration of exposure was short.

In 1969, Holden (186) reported one case of carcinoma of the prostate and one case of carcinoma of the bronchus, among 42 men exposed to cadmium fumes from 2 to 40 years. The total number of deaths was not reported. Morgan (316) found significantly higher cadmium concentrations in both renal and hepatic tissue taken from patients dying of bronchogenic carcinoma than in those taken from patients with other forms of cancer. Since cadmium is present in cigarette smoke and since smoking is associated with lung cancer, the association found by Morgan could be expected.

Kolonel (243) reviewed malignancies obtained from the Roswell Park Memorial Institute records covering the period from 1957 to 1964 for white males in the age range 50-79, in order to assess prior exposure to cadmium, particularly occupational exposure. The results suggest a significant association of renal cancer with exposure to cadmium and favour a synergistic effect between occupational exposure and smoking. The relative risk for men who both smoked and worked in high-risk occupations (e.g. electroplating, alloy-making, welding, manufacture of storage batteries) was more than four times that for men who did neither.

Lemen et al (262) have recently undertaken a retrospective cohort mortality study of 283 male workers who have been employed in a cadmium plant for a minimum of three years. The study indicates that there were greater total malignancies (25 vs. 15.9 expected), lung cancer (12 vs. 4.64 expected) and prostatic cancer (4 vs. 1.01 expected) among cadmium smelter workers than would be expected in the general population. The authors conclude that their observations strongly implicate cadmium exposure as a cause of certain types of malignant disease.

In summary, cadmium can certainly induce cancer in animals, mostly at sites of injections.

Some surveys among workers suggest that cadmium could also be carcinogenic in man but up to now the number of workers involved in each survey is too small to allow a definite conclusion to be drawn. No data (e.g. from Japan) is available to suggest that non-occupational exposure to cadmium constitutes a carcinogenic hazard.

VI - 4   MUTAGENICITY

There are conflicting reports in the literature concerning the mutagenic
action of cadmium compounds.   Shiraishi et al (446) reported that cadmium
sulphide induced chromosomal aberrations in culture human leukocytes (0.062
μg/ml culture medium) while Paton and Allison (365) could not find such an
effect of cadmium chloride (concentrations 5 x 10^{-8} and 3 x 10^{-9} moles/litre).
Röhr and Bauchinger (413) have found that higher concentrations of cadmium can
affect the mitotic apparatus as well as the chromosome structure of Chinese
hamster cells.   The mitotic activity has almost disappeared at 10^{-5} moles/
litre.   In 1972, Shiraishi and Yosida (445) reported an increased frequency
of chromosome abnormalities in peripheral leukocytes of seven patients with
Itai-Itai disease.   This observation could not be confirmed by a recent
investigation performed by The-Hung Bui et al (487) on five cadmium exposed
Swedish workers and four Japanese Itai-Itai patients.

Chromosome analyses in peripheral lymphocates of 24 workers of a zinc smelting
plant with increased blood levels of cadmium and lead were carried out by
Bauchinger et al (25).   The number of cells with structural chromosome aberrations was significantly increased as compared with 15 controls.   The observed
chromosome damage was mainly of the chromatid type (single break and exchanges)
accompanied by acentric fragments.   The mean of structural chromosome aberrations in the exposed group is increased by a factor of 3 (1.35 $\pm$0.99 in the
exposed group vs. 0.47 $\pm$ 0.92% in the controls).   Workers from a cadmium
plant in Belgium had a higher yield of severe chromosome anomalies (chromatid
exchange, disturbance of spiralization, chromosome translocation, and dicen-
tric chromosomes) when compared with unexposed workers (83).   However, the
observed chromosome aberrations reported by the last two groups of workers
cannot be causally related to cadmium because the workers were also exposed to
lead and a synergistic effect between the two metals is therefore possible.
No important chromosome aberrations were recorded in the lymphocyte cultures
of cattle intoxicated with heavy metals such as cadmium, lead and zinc (267)
and no significant genetic effects were obtained in mice given cadmium chloride (up to 3 mg/kg IP) (143).   Cadmium was not a mutagen in the mouse by the
dominant lethal assay (101, 143, 473).

VI - 5   EFFECT ON REPRODUCTIVE PROCESSES (FERTILITY, EMBRYOTOXICITY,

PERINATOLOGY)

Reproductive toxicology, which deals with the detection of undesirable effects
of agents on all reproductive processes, can be subdivided into three areas of
research: fertility, embryotoxicity and perinatology.

In fertility studies emphasis is placed on the detection of reversible and irreversible effects of agents on the fertility of several generations.   It has
been found that in the presence of 0.02 mM cadmium (as sulphate) human spermatozoa are immobilized (528).   The acute effect of cadmium on the testis has
been discussed in a previous chapter.   In mice no morphological changes have
been found in the testes after repeated administration of cadmium sufficient
to cause kidney damage (340, 382).   Suter (473) found that single intraperitoneal injection of 2 mg/kg $CdCl_2$ to female mice did not modify their reproductive capacity.   Nordberg (346) exposed male CBA-mice to cadmium by subcutaneous injection of 0.4 mg $CdCl_2$/kg body weight for 5 days/week for 6 months.
A decrease in normal (testosterone-dependent) proteinuria was shown and morphological examination of the seminal vesicles revealed a smaller weight and
size as well as histological indication of lower secretory activity of the
epithelium compared to controls.   A decreased testosterone activity in cadmium treated animals is compatible with this finding.

No specific effect of cadmium on testes has been reported after long term
exposure of man to cadmium, but this may be due to a lack of studies.   Favino
et al (110) found one case of impotency with low urinary excretion of testosterone among 10 cadmium workers.   The study is too limited to draw any
conclusion but stresses the interest of further investigating the testicular
function in persons exposed to cadmium.

In embryotoxicity studies emphasis is placed on the detection of adverse effects of agents on the offspring of animals treated during pregnancy.   It is
known (59, 361, 362, 363) that cadmium can induce embryolethality in pregnant
rats.   Parizek (362, 363) showed that after maternal cadmium injections of 2-
4 mg/kg the foetal placenta could be fairly rapidly damaged, leading to death
of the embryos in utero.

Chernoff (59) showed that in rats daily subcutaneous injections of 4 mg $CdCl_2$/ kg body weight from day 13 to 19 of pregnancy caused also embryolethality; at doses of 12 mg $CdCl_2$/kg body weight 50% of all embryos died in utero. Chernoff (59) also demonstrated that at increasing dose levels of cadmium foetal body weight decreased. Ishizu et al (558) demonstrated that one single subcutaneous injection of 0.33mg/kg body weight cadmium chloride on day 7 of pregnancy induced embryolethal effects in mice, teratogenic effects were seen above 0.63mg/kg. In rats, Fabricius et al (556) showed that one single intraperitoneal injection of 2.5 mg/kg body weight of cadmium chloride given early in pregnancy also produced teratogenic effects. However, lower doses were not tested. Besides embryolethal and teratogenic effects induced by cadmium in rats and mice, cadmium could also induce teratogenic effects in hamsters, and frogs (22, 59, 60, 64, 111, 115, 118, 140, 221, 420, 438). In mice an intraperitoneal injection of 5 mg $CdSO_4$/kg body weight on day 7 of pregnancy induced exencephaly in 40-50% of the offspring. Pretreatment of 1 mg $CdSO_4$/kg on day 6 of pregnancy could inhibit the teratogenic effect of cadmium in mice (438). In frog-embryos, maintained in water containing 2 mg $CdSO_4$ per litre, also failure of closure of the neural tube (leading to exencephaly) was seen: at increasing doses of cadmium the number of embryos displaying exencephaly increased rapidly (221). If cadmium was given to pregnant mice together with zinc the embryolethal and teratogenic effects produced by intraperitoneal administration of 5 mg $CdSO_4$/kg, disappeared (438). The same effect was seen with the use of selenium (112, 113, 187, 363) but not with cobalt (438). The teratogenic effects induced by cadmium in rats were of a different nature. The rate of anomalies rose from 4% at 4 mg $CdCl_2$/kg body weight (subcutaneous) to 70% at 12 mg $CdCl_2$/kg body weight (subcutaneous). The main anomalies which were induced by cadmium in the rat offspring were: micrognathia, club foot, cleft palate and small lungs (59).

Pond and Walker (388) determined the effects of the addition of 200 ppm cadmium (as $CdCl_2$) to the diet of pregnant rats. Number of live or stillborn pups per litter was not significantly affected by diet but cadmium significantly reduced pup birth weight. No grossly abnormal pups were noted. The prenatal toxicity of cadmium chloride was also tested on mice by Zeller and Peh (548). The compound was administered orally from day 6 to 15 of pregnancy at the following doses: 4.27, 12.8 and 38.4 mg/kg which corresponded to 1/45, 1/15 and 1/5 of the oral LD_{50} in the species investigated. The lowest dose had no teratogenic effect. At 12.8 mg/kg slight toxic effects were observed in the mothers but no teratogenic action was evidenced. The highest dose was clearly teratogenic and embryolethal. A similar study was performed by Machemer on rats which were given 0, 3, 10, 30, 100 mg/kg $CdCl_2$ orally from day 6 to 16 of pregnancy. At 30 mg/kg per day death of some mothers occurred and at 10 mg/kg per day the body weight increase of the mothers was significantly depressed.

In the 30 mg/kg group foetal malformations were found. The no-effect level was estimated at 3 mg/kg p.o. per day for the mothers and at 10 mg/kg p.o. per day for the foetuses. A classical three generation study (6 matings) was performed in rats which were given cadmium in their diet (as $CdCl_2$) at the following doses: 1, 3, 10 and 100 ppm. The fertility index, the number of pups per litter and the weaning index were not influenced by cadmium administration. In the hihgest dose group the weight of the pups was slightly reduced. The results of the histopathological examination of the mothers and newborns (F3B generation) are not yet available (Loser and Lorke, personal communication).

Prigge et al (393) exposed pregnant and non pregnant rats to $CdCl_2$ aerosol (0.2 mg/m^3) alone or in combination with 250 ppm CO for 20 days. They found that exposure to combination of Cd and CO as well as exposure to Cd alone produced higher Cd concentration in the liver of non-pregnant rats than in the respective groups of pregnant animals.

Dencker (87) has investigated the accumulation of cadmium administered intravenously (0.1 to 0.21 μg Cd^{2+} per animal) in embryonic (foetal) and placental structures and also in the whole maternal genital apparatus in mice and golden hamsters at different stages of gestation. Cadmium administered on the 8th day accumulated in the primitive gut of the embryos. No cadmium was detected in the embryos after administration on or after the 9th day (hamster) and 11th day (mouse). Cadmium is also heavily accumulated in the decidua, the yolk sac, the ectoplacental cone and later in the chorioallantoic placenta. A high ovarian accumulation (corpus luteum and follicles) was also found. The author summarized the significance of his findings as follows:

1) the heavy accumulation in the embryonic gut wall in early gestation makes probable a direct effect on the embryo;

2) the decidual and placental uptake indicates that the maternal-embryonic relationship, and, hence embryonic nutrition, may be disturbed;

3) the ovarian accumulation may also disturb steroid synthesis leading to an interruption of pregnancy.

Very little is known about the teratogenic effects of cadmium in humans. Friberg et al (129) quote Cvetkova (1970) for a study of 106 women employed in cadmium industries and 20 control women. The weights of 27 children born to exposed women were significantly lower than the weights of children born to control women. In 4 of the 17 children born to women in the zinc smeltery group, signs of rickets and delayed development of teeth were observed.

In perinatology studies emphasis is placed on the determination whether an agent, the administration of which is begun shortly before birth and continued into the period of lactation, has undesirable effects on, for example, lactation and postnatal development of the suckling young. No such perinatology studies with cadmium are known.

Chapter VII

SOME EXISTING GUIDES AND STANDARDS FOR ENVIRONMENTAL LEVELS

VII - 1   FOOD

An FAO/WHO expert committee has proposed a provisional maximum tolerable
weekly intake of 400-500μg per individual (355) which would correspond to an
average daily intake of 57.1-71.4μg (i.e. ±1μg/kg).   This estimate is in fact
based on Friberg's model.   It is believed that if the intake does not exceed
this level it is unlikely that the level of cadmium in the renal cortex will
exceed 50ppm.   Accordingly, as the amount of food consumed by an adult person
is about 10kg per week the mean content of cadmium in food should not exceed
0.04-0.05ppm (99).

The Japanese Ministry of Health and Welfare have proposed a level of 300μg per
day as the maximum acceptable dietary intake for cadmium.   Two other guidelines equivalent to the 300μg cadmium per day which not be exceeded were also
established by the Ministry:

1.   a daily urinary concentration of 9μg/l and
2.   an unhulled rice content of 0.4 ppm.

Following the outbreak of Itai-Itai disease, the Japanese Ministry of Health
and Welfare recommended in 1970 that farmers in several polluted areas should
not eat rice containing over 1.0ppm cadmium for unpolished rice and over
0.9ppm cadmium for polished rice.   The provisional guidelines for starting
detailed investigations of cadmium environmental pollution are 0.4ppm in rice
or concentrations in drinking water higher than 10μg/l.

Reference values for Cd concentrations in various food items have been published by the Germany Ministry of Health in 1974, i.e. meat: 0.008ppm;
vegetable: 0.1ppm;   fruit: 0.05ppm).

VII - 2   DRINKING WATER

The tentative upper limit for cadmium in drinking water was initially set by WHO at 10μg/l (538) but a recent WHO working group (539) has recommended that cadmium concentration in water should not exceed 5μg/l.   The Commission of the European Communities has proposed the same standard (72).

In Japan, USA and USSR the environmental water quality standard for cadmium is 10μg/l, and in Germany: 6μg Cd/litre.

VII - 3   AIR

3.1 General Environment

In Japan the provisional guideline for cadmium in ambient air is 0.88 - $2.92μg/m^3$.   The lower guideline value (0.88) is that which should usually be maintained whereas the higher value should not be exceeded even when the weather conditions are most disadvantageous.

3.2 Industrial Environment

In the USA, ACGIH (1976) has proposed a maximum allowable concentration for cadmium oxide fumes of $0.05mg/m^3$, and a time weighted average (TLV or TWA) for cadmium dust of $0.05mg/m^3$.   NIOSH have recommended a TLV of $40μg/m^3$ and a 15 minute ceiling limit of $200μg/m^3$ for both dust and fumes.

In the USSR, the TLV for cadmium fume is $0.1mg/m^3$.

The TLV for cadmium dust and fume is $10μg/m^3$ in Finland and $20μg/m^3$ in Sweden.

In Germany the MAK for cadmium oxide is $0.1mg/m^3$ (244).

In the UK the TLVs proposed by the Health and Safety Executive, (technical data note 2/75) are:

- 0.2 mg/m^3 for cadmium as metal dust or soluble salt
- 0.05mg/m^3 for cadmium oxide fumes.

In Switzerland the Caisse Nationale Suisse en cas d'accidents 1976 has proposed a TLV of 0.2mg/m^3 for cadmium dust.

VII - 4   SLUDGE

Because of the risk of cadmium uptake by crops a maximum addition of 1 ton (dry weight) per ha of sludge (containing 5 to 15ppm cadmium dry weight) has been recommended in Sweden (89).

VII - 5   EMISSION STANDARDS

In Japan, the liquid effluent standard for cadmium is 0.1ppm (assuming a minimal in-river dilution factor of 10) and the atmospheric emission standard has been set at 1.0mg/Nm^3 (349).   In Germany, the technical instruction "Air" (Technische Anleitung zur Reinhaltung der Luft August 1974) recommends a maximum emission standard for cadmium of 20mg/m^3.   In France the liquid effluent standard is 3mg/litre.

VII - 6   FOOD CONTAINERS

The national regulations concerning the amount of cadmium released from ceramics have been summarized by Engberg and Bro-Rasmussen (98) and are presented in the following table:

National regulations concerning cadmium from ceramics
(from references 98 and 99)

Country	Extraction conditions	Cadmium limit
USA ceramics, enamelware	4% acetic acid 24h at room temperature	0.5 mg/l
SWEDEN	4% acetic acid 24h at room temperature	0.1 mg/l
UK glazed ceramic ware enamel (cooking ware) hollow ware kitchen utensils	4% acetic acid 2h at 120°C 22h cooling period	0.7 mg/l 0.2 mg/l 5.0 mg/m^2
DENMARK ceramics, glass & enamelware	4% acetic acid 90min. at 100°C	1.0 mg/l
IRELAND ceramic products	4% acetic acid 24h at room temperature	0.5 mg/l
ITALY (proposal) ceramic products ceramic enamels	3% acetic acid 3 x 24h at 40°C	0.5mg/l 1st attack 0.2mg/l 3rd attack
NETHERLANDS ceramic products ceramic enamels	3% acetic acid 3 x 2h at 40°C 3 x 2h at 100°C	0.02 mg/dm^2 (3rd attack tableware) 0.02mg/dm^2 (3rd attack cookingware
Proposed EEC Directive ceramic products	4% acetic acid 24h at room temperature	0.1mg/dm^2) flatware) table- 0.5mg/dm^2) ware hollowware) 0.25mg/l child's plate 0.05mg/dm^2 cookingware, flat 0.25mg/l hollow ware

In South Africa, the maximum concentration of cadmium allowed in any type of food container is 1ppm (347).

Chapter VIII

NEEDS FOR FURTHER RESEARCH

Only research needs relevant to public health will be considered.

1. Cadmium analysis

The analytical methods for determination of cadmium at concentrations found in biological material should be critically evaluated and if possible standardized. This constitutes a prerequisite for the evaluation of the significance of cadmium level in blood and in urine, the development of a biological quality guide (as it was possible for lead (549)) and the monitoring of population groups.

Monitoring of the environment (air, water, dust, soil) in Europe should be continued to better assess the overall exposure of the population and pinpoint groups at risk.

2. Cadmium metabolism

It is evident that much less is known about cadmium metabolism than about lead or mercury. Studies of cadmium absorption, distribution, excretion and accumulation in man (of different ages) and in animals are required in order to:

a. determine the absorption rate of cadmium administered by different routes and its excretion rate,

b. evaluate total body burden of different population groups in Europe,

c. discover methods of biological monitoring (relation between exposure and/or body burden and cadmium concentration in biological media),

d. determine whether the chemical form of cadmium influences its fate in the organism,

e. determine factors which modify cadmium metabolism (age, diet, zinc, vitamins, disease status ...).

In this regard one can mention some clinical studies which would be most useful:

1. follow cadmium in urine and in blood with time, in newly cadmium exposed workers at the same time as exposure is evaluated by personal monitoring,

2. cadmium determination in human tissues of different age groups,

3. balance studies on volunteers.

3. Toxicity (experimental and epidemiological studies)

Research activities are needed in different areas:

a. It is important to better define the critical level of cadmium concentration in kidney for causing kidney damage in animals and in humans (autopsy of persons with different exposures to cadmium: workers, persons living around emission sources ...). In this connection it would be useful to clarify whether it is the overall cadmium level which relates to toxicity or a specific fraction. The knowledge of the critical level of cadmium in the kidney together with information regarding its metabolism would help estimating the acceptable daily intake of cadmium.

b. Because of the conflicting reports in the literature, epidemiological research should be performed on workers to confirm or refute the carcinogenic and mutagenic action of cadmium in man (e.g. large scale mortality studies of workers exposed to cadmium, international collaboration is necessary because only small groups are available in each country).

c. Epidemiological studies on differently exposed groups should be carried out to investigate the relationship between hypertension and cadmium exposure. For such a kind of survey a biological method of monitoring would be most useful to distinguish between different exposures and/or body burdens.

d. With the use of sensitive biological parameters (e.g. $\beta_2$microglobulin in urine ...) potentially high exposure groups in the general population (e.g. old women living around cadmium emitting sources) should be investigated to find out whether (in addition to cadmium workers) critical human groups can already be found in Europe. The total exposure of the groups investigated should also be carefully evaluated (cadmium in air, in diet, in water).

e. Cross-sectional and longitudinal studies of workers exposed to cadmium (evaluation of airborne exposure and monitoring of various biological and physiological parameters, e.g. Cd-U, Cd-B, lung function tests, kidney function) are required to estimate the no-response level of cadmium in air.

f. Further studies of long term effect of cadmium on testicular function in man (hormonal status of workers) are relevant in view of the animal findings.

g. Long term exposure of animals to cadmium dust would be indicated to confirm (with the use of sufficiently sensitive tests), the no-effect level of cadmium on the lungs. Since local action on lung macrophages has been suggested, it would be interesting to determine whether cadmium can interact with the lung defence mechanism (e.g. susceptibility to infection).

h. Long term exposure of animals to cadmium added to their diet would permit the evaluation of the no-effect level for functional kidney changes and to study the relationship between exposure, biological changes and cadmium levels in blood, urine, and tissues (it has been suggested that cadmium absorption may be different when administered in the diet or in drinking water).

i. The carcinogenic activity of cadmium administered by inhalation to animals should be investigated. If it appears that the carcinogenic activity of cadmium depends on the route of administration, the mechanism responsible for the different response should then be studied (e.g. protective role of liver metallothionein when cadmium is administered orally ...).

j. The study of the mechanism of cadmium hypertension in animals (role of zinc level, renin system ...) may help clarify the risk for man.

k. The study of interactions of cadmium and other cancerogens (e.g. benzpyren) in animals might be useful to pinpoint the potential risk for man.

l. The role of metallothionein in cadmium toxicity should be further investigated (transport, synergistic action, defence mechanism ...).

m. The risk of cadmium for the human foetus should be assessed. Further studies on the transplacental transfer of cadmium are indicated.

Chapter IX

TENTATIVE PROPOSAL FOR A NO-EFFECT LEVEL FOR LONG TERM EXPOSURE TO CADMIUM

It is currently estimated that the critical organ (i.e. the organ in which the first adverse effect occurs following long term exposure to cadmium) is the kidney. The proposal for no-effect levels must therefore be based principally on the evaluation of the risk of kidney impairment by cadmium.

Since no easily accessible biological parameter (blood, urine) has definitely been proved to reflect either body burden or exposure, a biological quality guide for cadmium cannot yet be elaborated.

A "classical" approach - i.e. the finding of "no-effect" or "no-response" exposures - which takes into consideration the various routes of entry must therefore be considered in order to protect public health.

When progress is made regarding the evaluation of internal load and/or exposure by means of biological parameters, and when the relationships between these parameters and responses are identified, a "biological quality guide" can possibly be elaborated which is certainly needed to detect potential groups at risk. Currently two different approaches can be considered to estimate the environmental hazard of cadmium to man.

The first one is based on the direct evaluation of the no-effect exposures or the no-response exposures of cadmium administered by different routes to man and to animals and a comparison of these values with the current exposure of man. Unfortunately sufficient epidemiological and experimental data are not available to establish precise dose-response curves for the adverse effects of cadmium. Furthermore published data have usually not been presented in a manner which permits the elaboration of dose-response curves.

The second one which has been proposed by Friberg for the estimation of no-effect or no-response levels following long term exposure to cadmium is based on the assumption that there is a relationship between exposure and the concentration in the target organ (i.e. the kidney cortex).

125

If one knows the critical concentration in this organ, it is possible by the
use of a simple model to calculate the daily intake required to reach this
level in a stated length of time.

1.   First Approach: evaluation of no-effect levels for long term cadmium
                       effect on kidney.

Experimental data on animals

Oral administration

On a purely scientific basis we must recognize that the no-effect level
of cadmium administered orally to animals is still unknown.   Only an
"oriented" guess can be formulated in confronting the various reports.

I would suggest that until Fowler's observations on renal vessels in cad-
mium-exposed rats have been confirmed by other investigators, to
consider 1 ppm Cd (in drinking water) as a tentative no-effect level for
rats.   Assuming that a 200g rat drinks 25ml water per day, 1 ppm Cd in
drinking water would correspond to an oral daily intake of 125µg/kg.

Thus, applying a safety factor of 100 and neglecting concomitant ambient air exposure, the acceptable oral daily intake for an
adult man (60kg) would be 75µg/day.

Inhalation

Friberg found that 5mg Cd/m^3 administered to rabbits (3 hours/day) cause
proteinuria after 4 months.   No long term studies with the use of sensi
tive methods have been performed to evaluate the no-effect level of airborne cadmium on the kidney of animals.

Epidemiological data

Inhalation

The few studies performed among workers exposed to cadmium suggest that
an exposure to a "respirable" cadmium dust concentration above 20µg/m^3

(8 hours/day for 20 years) could still induce kidney lesions after a 20 year exposure. If it is extrapolated to a continuous exposure (24 h/day for 70 years) this concentration has to be reduced to about $2\mu g/m^3$ (dust 5μ). Thus the no-effect level for the kidney is probably below this value.

Oral intake

Epidemiological surveys performed in Japan suggest that a continuous oral daily intake exceeding 200μg cadmium could cause an increased prevalence of kidney damage in persons over 50 years (although the amount of cadmium absorbed through the lung by the Japanese living in polluted areas is not known, it is believed to be negligible in comparison with oral absorption). It has been estimated that such an intake corresponds to an average urinary excretion of 6μg cadmium/litre but it should be stressed that this value is highly tentative and more research work is required before proposing a biological threshold for cadmium in urine.

Summary of the first approach

Since the kidney constitutes the critical organ, the evaluation of the no-effect level of cadmium on this organ is of paramount importance. For the general population, the major source of exposure to cadmium is through food consumption. On the basis of the available human data, when uptake by inhalation is negligible (non smoker living in urban or rural areas), the threshold effect level of cadmium by ingestion is around 200 μg daily corresponding to an absorption of 12μg/day (absorption rate estimated at 6%). In smokers, the threshold effect level for oral absorption is reduced by about 1.9μg (i.e. to 10.1μg corresponding to an oral intake of 169μg). When exposure by inhalation is significant the threshold effect level by ingestion must be reduced proportionally to the amount (x) absorbed by inhalation (i.e. $\frac{x\ 100}{6}$). In some very special circumstances, however, the amount absorbed by inhalation alone (17.9μg/day) could exceed the total daily absorption (12μg) (see Chapter V) corresponding to the threshold effect level.

2.   Second Approach

Friberg and his collaborators have proposed a model to evaluate the long
term accumulation of cadmium in humans.   This model is based on the following assumptions:

-      one third of the body burden is in the kidneys (concentration in
       cortex 50% higher than the renal mean value),
-      of the absorbed cadmium 10% is considered to be rapidly excreted and
       this fraction is excluded from the calculation because it ultimately
       could contribute less than 1% of the total body burden of cadmium,
-      4.5% of ingested cadmium and 25% of inhaled cadmium are absorbed.
The model can take into consideration variation of kidney weight and
calorie intake with age.   The model permits the calculation of the cadmium intake by different routes which is necessary to reach a critical
concentration in kidney cortex.   The basic information required to apply
this model are the critical concentration of cadmium in the kidney cortex
and the excretion rate of cadmium in humans.   Friberg and his collaborators estimated that the excretion rate is in the order of 0.005% or less.
This value gives the most plausible intake figure to reach a renal cortex
concentration of 50µg/g wet weight at age 50 (value which has been found
by analysis of autopsy samples).   The critical concentration of cadmium
in the kidney cortex is difficult to evaluate because the human and experimental data available are very limited.   These data can be briefly
summarized as follows:

Animal data

Oral administration

A few investigators have determined cadmium concentration in the kidney
of animals with signs of renal disturbances.   In rats and in rabbits
renal cortex concentration of approximately 135 and 200µg/g wet weight
respectively have been reported to be associated with signs of kidney
damages.   In rats on a calcium deficient diet a concentration of 90µg/g
has been found when increased enzymuria was present.

Recently Fowler et al (125) have indicated that cadmium concentrations in
rat kidney even below 1µg/g can already be associated with a constriction

of smaller renal arteries and Koller et al (241) have also found that in mice kidney concentration below $3\mu g/g$ is also associated with tubular lesion. However, other authors have not found renal change in rats with a cadmium concentration in whole kidney of 12ppm (i.e. ±18ppm in renal cortex). These results are currently too contradictory to suggest the critical cadmium concentration in kidney of rats.

Parenteral administration

After parenteral administration of cadmium, renal cortex concentration of $330-480\mu g/g$ wet weight in rats and $250\mu g/g$ in rabbits have been measured in animals with kidney lesion.

Human data

If one makes several assumptions regarding deposition, absorption and distribution of cadmium in the organism following its inhalation one can, on the basis of the results of two recent epidemiological studies on workers, estimate that a renal cortex concentration of about $300\mu g/g$ wet weight can be associated with function signs of kidney lesion. After oral absorption of cadmium, one can also estimate that a kidney cortex concentration of $364-370\mu g/g$ wet weight could increase the prevalence of proteinuria which is not very different from the above estimation based on the examination of workers exposed by inhalation.

The Subcommittee on the Toxicology of Metals under the Permanent Commission and International Association of Occupational Health has recently endorsed Friberg's proposal of $200\mu g$ cadmium/g wet weight as the tentative critical concentration in human kidney cortex (344). However, the tentative character of this proposal should be stressed because it is based on few observations on humans and some experimental results suggest that kidney impairment could occur at a lower kidney concentration. Considering first the oral intake and neglecting intake via ambient air or smoking, Friberg calculated that the necessary daily cadmium intake to reach the critical concentration of 200 ppm in the kidney cortex at age 50 is $248\mu g$. It can also be estimated that the adult oral intake of cadmium necessary to reach 50ppm in renal cortex at age 50 is $62\mu g/day$. Considering only the respiratory intake, the necessary cadmium concentration in ambient air to reach the critical cadmium concentration in kidney cortex at age 50 is $2\mu g/m^3$ (ventilation $20m^3/24$ h).

On the basis of the same model the necessary daily cadmium absorption (by all routes) to reach 200ppm in the renal cortex at age 50 is $10\mu g$. However, as Fleisher et al have recently pointed out, although the assumptions made seem valid as working hypotheses, there are many uncertainties in this model:

1. The tentative value of the critical level in the target organ has already been stressed. (Some animal experiments suggest that the critical level is much lower than 200 ppm).
2. The unusual nature of the cadmium binding protein in the kidney makes it uncertain whether overall cadmium levels relate to toxicity.
3. Analytically valid long term human balance studies are not available to precisely estimate the rate of cadmium absorption and excretion.

3. Conclusions

The two approaches give the following estimates of threshold effect levels of cadmium for adult human beings:

1. Oral exposure (considering pulmonary absorption as negligible)
    First approach: threshold effect exposure: $200\mu g/daily$
    Second approach: threshold effect exposure: $248\mu g/daily$

2. Pulmonary exposure (considering oral intake as negligible)
    First approach: threshold effect exposure: $12\mu g/m^3$
    Second approach: threshold effect exposure: $2\mu g/m^3$

3. Combined exposure
    For the general population the diet constitutes the main source of exposure to cadmium and therefore oral exposure can never be neglected. In circumstances where the pulmonary exposure is not negligible the threshold effect oral intake must be reduced by the following amount: $\dfrac{x \quad 30 \quad 0.16}{0.06}$

    where x = airborne cadmium concentration in $\mu g/m^3$
              30   = number of m^3 inhaled daily by an adult
             0.16 = absorption rate through the lung
             0.06 = absorption rate through the gastrointestinal tract.

A daily intake of 200µg corresponds to a cadmium concentration in the basic foodstuff of 0.25ppm (assuming half the cadmium intake comes from this foodstuff and 400g of this foodstuff are ingested daily).

An attempt has been made to compare the threshold effect levels estimated by both approaches described above with the estimated current exposure of the general population in order to estimate whether a risk could already exist for some population groups (excluding occupationally exposed workers). This comparison is summarized in the following table. It appears that population groups living very close to a Cd emission source could be at risk. However because of the great uncertainties in the estimate of both current exposure and threshold effect level this conclusion should be regarded only as a working hypothesis. Research work is urgently needed to test it.

Evaluation of the safety factors by comparing the
tentative threshold effect levels (absorption)
and the estimated current amounts of cadmium absorbed

Current absorption (Table 11) (µg/day)	Safety Factor	
	First approach	Second approach
Rural area		
*N.S. = 0.36 - 5.78	33.3 - 2.1	27.8 - 1.7
** S. = 2.28 - 7.70	5.3 - 1.6	4.3 - 1.3
Urban area		
*N.S. = 0.37 - 7.88	32.4 - 1.5	27 - 1.3
** S. = 2.29 - 9.8	5.2 - 1.2	4.4 - 1.0
Industrial area		
*N.S. : 0.39 - 21.64	30.8 - 0.6	25.6 - 0.5
** S. : 2.31 - 23.56	5.2 - 0.5	4.3 - 0.4
Threshold effect level (absorption)	12µg/day	10µg/day

*N.S.:   non smoker ** S.:   smoker

REFERENCES

1. ABDULLAH M.I. and ROYLE L.G.
    Cadmium in some british coastal and fresh water environments in
    European Colloquium Problems of the Contamination of man and his
    environment by mercury and cadmium.
    Luxembourg, July 1973.

2. ADAMS R.G. and CRABTREE N.
    Anosmia in alkaline battery workers.
    Brit. J. Ind. Med. 18, 216, 1966.

3. ADAMS R.G., HARRISON J.F. and SCOTT P.
    The development of cadmium-induced proteinuria, impaired renal
    function and osteamalacia in alkaline battery workers.
    O.J. Med. 38, 425, 1969.

4. AHLMARK A., AXELSSON B., FRIBERG L. and PISCATOR M.
    Further investigations into kidney function and proteinuria in
    chronic cadmium poisoning.
    Int. Congr. Occup. Health 13, 201, 1961.

5. ALLWRIGT S.P.A., COULSON A., DETELS R. and PORTER C.E.
    Mortality and water-hardness in three matched communities in
    Los Angeles.
    The Lancet, October, 860, 1974.

6. AMACHER D.E., EWING K.L.
    A soluble cadmium-binding component in rat and dog spleen.
    Arch. Environ. Health 30, 510, 1975a.

7. AMACHER D.E., EWING K.L.
    Cadmium deposition in canine heart and major arteries following
    intravascular administration of cadmium chloride.
    Bull. Environ. Cont. Toxicol. 14, 457, 1975b.

8. ANDREUZZI P. and ODESCALCHI C.P.
    Intossicazione sperimentale acuta da chlorure di cadmio nel conig-
    lio I modificazioni dell'attivita transaminasica glutammico-ossa-
    lacetica nel siero.
    Boll. Soc. Ital. Biol. Sper. 34, 1376, 1958 (in Italian).

9. ANKE M. and SCHNEIDER M.J.
    Der Zink-, Kadmium und Kupferstoffwechsel des Menschen.
    Arch. Veterenärmed 25, 805, 1971.

10. ANWAR R., LANGHAM R., HOPPER C.A., ALFREDSON B.W. and BYERRUM R.V.
    Chronic toxicity studies. III Chronic Toxicity of cadmium and
    chromium in dogs.
    Arch. Environ. Health 3, 456, 1961.

11. ARENA J.M.
    Poisoning chemistry-symptoms - treatment.
    Thomas, Springfield, 111, 1963.

12. AUBERT M. et DONNIER B.
    Pollution du milieu marin par le mercure et le cadmium en
    Méditerranée.
    In European Colloquium. Problems of the contamination of man and
    his environment by mercury and cadmium.
    Luxembourg, July 1973.

13. AXELSSON B. and PISCATOR M.
    Renal damage after prolonged exposure to cadmium.
    Arch. Environ. Health 12, 360, 1966 .

14. AXELSSON B., DAHLGREN S.E. and PISCATOR M.
    Renal lesions in the rabbit after long-term exposure to cadmium.
    Arch. Environ. Health 17, 24, 1968.

15. BAADER E.W.
    Die chronische Kadmiumvergiftung.
    Dtsch. Med. Wochenschr. 76, 484, 1951.

16. BAADER E.W.
    Chronic cadmium poisoning.
    Ind. Med. Surg. 21, 427, 1952.

17. BAEJER H.P.
    Death due to cadmium oxide fumes.
    Indust. Med. Surg. 363, May 1969.

18. BAGLAN R.J., BRILL A.B., SCHULERT A., WILSON D., LARSEN K., DYER N., MANSOUR H., SCHAFENER W., HOFFMAN L. and DAVIES J.
    Utility of placental tissue as an indicator of trace element exposure to adult and fetus.
    Environ. Res. 8, 64, 1974.

19. BANIS R.J., POND W.G., WALKER E.F. and O'CONNOR J.R.
    Dietary cadmium, iron and zinc interactions in the growing rat.
    Proc. Soc. Exp. Biol. Med. 130, 802, 1969.

20. BARBIERI L., COLOMBI R. and STRANEO G.
    Modificazioni istologiche delle isole pancreatiche del coniglio-dopo intossicazione cronica da cadmio.
    Folio Med. 44, 1120, 1961.

21. BAROWICZ T., JASTRZEBSKI M., WOJCIK K.
    Early cadmium induced changes of the permeability of testicular blood vessels to ^{131}I albumin in the rat.
    Bull. Acad. Pol. Sci. Cl. 1-22, 201, 1974.

22. BARR M.
    The teratogenicity of cadmium chloride in two stocks of Wistar rats.
    Teratology 7, 237, 1973.

23. BARRETT H.M., IRWIN D.A. and SEMMONS E.,
    Studies on the toxicity of inhaled cadmium.
    I. The acute toxicity of cadmium oxide by inhalation.
    J. Ind. Hyg. Toxicol. 29, 279, 1947a.

24. BARRETT H.M. and CARD B.Y.
    Studies on the toxicity of inhaled cadmium.
    II. The acute lethal dose of cadmium oxide for man.
    J. Ind. Hyg. Toxicol. 29, 286, 1947b.

25. BAUCHINGER M., SCHMID E., EINBRODT H.J. and DRESP J.
    Chromosome aberrations in lymphocytes after occupational lead and cadmium exposure.
    Mutation Research 40, 57, 1976.

26. BAUMERT H.P., MUHLE H., OTTO F. and STUER D.
    The effect of heavy metal inhalation on cell number and metabolism
    of alveolar macrophages of the mammal lung.
    Report to CEC, Aug. 1976.

27. BAWDEN J.W. and HAMMARSTROM L.E.
    Distribution of cadmium in developing teeth and bone of young rats.
    Scand. J. Dent. Res. 83, 179, 1975.

28. BENTLEY P.J., YORIO T. and FLEISHER L.
    Effects of cadmium on the hydro-osmotic and natriferic responses
    of the toad bladder to vasopressin.
    J. Endocr. 66, 273, 1975.

29. BERGNER K.G., LANG B. and ACKERMANN H.
    Zum Cadmiumgehalt deutscher Weine.
    Mitt. Rebenwein Obstbau Fruechteverwert. 22, 101, 1972.

30. BERLIN M. and ULLBERG S.
    The fate of ^{109}Cd in the mouse. An autoradiographic study after
    a single intravenous injection of ^{109}CdCl$_2$.
    Arch. Environ. Health. 7, 686, 1963.

31. BERNARD A., ROELS H., HUBERMONT G., BUCHET J.P., MASSON P.L. and
    LAUWERYS R.
    Characterization of the proteinuria in cadmium exposed workers.
    Int. Arch. Occup. Environ. Health. 38, 19, 1976.

32. BETON D.C., ANDREWS G.S., DAVIES H.J., HOWELLS L. and SMITH G.F.
    Acute cadmium fume poisoning, five cases with one death from
    renal necrosis.
    Brit. J. Industr. Med. 23, 292, 1966.

33. BIERENBAUM M.L., FLEISCHMAN A.I., DUNN J. and ARNOLD J.
    Possible toxic water factor in coronary heart disease.
    The Lancet May 3, 1008, 1975.

34. BONNELL J.A.

   Emphysema and proteinuria in men casting copper-cadmium alloys.

   Brit. J. Ind. Med. 12, 181, 1955.

35. BONNELL J.A., KAZANTZIS G. and KING F.

   A follow-up study of men exposed to cadmium oxide fume.

   Brit. J. Indust. Med. 16, 135, 1959.

36. BONNELL J.A., ROSS J.H. and KING E.

   Renal lesions in experimental cadmium poisoning.

   Brit. J. Industr. Med. 17, 69, 1960.

37. BOPPEL B.

   Blei- und Cadmiumgehalte von Lebensmitteln.

   Z. Lebensm. Unters.-Forsch. 160, 299, 1976.

38. BOUISSOU H. et FABRE M.Th.

   Lésions provoquées par le sulfate de cadmium sur le testicule du rat.

   Arch. Mal. Prof. 26, 127, 1965.

39. BOUQUIAUX J.

   Mercury and cadmium in the environment, first results of an enquiry on a European scale. In European Colloquium: Problems of the contamination of man and his environment by mercury and cadmium.

   Luxembourg, July 1973.

40. BROWNING E.

   Toxicity of Industrial Metals.

   Butterworths, London 1961.

41. BRUNE D., FRYKBERG B., SAMSAHL K. and WESTER P.O.

   Determination of elements in normal and leukemic human whole blood by neutron activation analysis AE-60, Aktiebolaget Atomenergie. Stockholm cited by Friberg et al. (129).

42. BUHLER R.H.O. and KAGI J.H.R.

   Human hepatic metallothioneins.

   FEBS Letters, 39, 229, 1974

43. BLUMER F.M.R., ROTHWELL H.E. and FRANKISH E.R.
    Industrial cadmium poisoning.
    Can. Public Health J., 29, 19, 1938.

44. BUXTON R., St. J.
    Respiratory function in men casting cadmium alloys.
    II. The estimation of the total lung volume, its subdivision and
    the mixing coefficient.
    Brit. J. Ind. Med. 13, 36, 1956.

45. CAMERON E. and FOSTER C.L.
    Observations on the histological effects of sub-lethal doses of
    cadmium chloride in the rabbit.
    J. Anat. 97, 269, 1963.

46. CARLSON L.A. and FRIBERG L.
    The distribution of cadmium in blood after repeated exposure.
    Scand. J. Clin. Lab. Invest. 9, 67, 1957.

47. CARROLL R.E.
    The relationship of cadmium in the air to cardiovascular disease
    death rates.
    JAMA 198, 267, 1966.

48. CARTENSEN J. and POULSEN E.
    Public Health aspects of environmental pollution with mercury and
    cadmium in Scandinavia.
    In European Colloquium: Problems of contamination of man and his
    environment by mercury and cadmium. Luxembourg, 1973.

49. CASTANO P.
    Chronic intoxication of cadmium experimentally induced in rabbits.
    Path. Microbiol. 37, 280, 1971.

50. CAUJOLLE F., OUSTRIN J. et SULVE-MAMY G.
    Fixation et circulation entérohépatique du cadmium
    Europ. J. Toxicol. 4, 310, 1971.

51. CERNIK A.A. and SAYERS M.H.P.
    Application of blood cadmium determination to industry using a
    punched disc technique.
    Brit. J. Indust. Med. 32, 155, 1975.

52. CHAUBE S., NISHIMURA H., SWINYARD C.A.
    Zinc and cadmium in normal human embryos and fetuses.
    Arch. Environ. Health 26, 237, 1973.

53. CHEN R.W., HOEKSTRA W.G. and GANTHER H.E.
    An unstable Cd-binding protein in the soluble fraction of rat testes.
    Fed. Proc. 32, 929, 1973.

54. CHEN R.W., WAGNER P.A., HOEHSTRA W.G. and GANTHER H.E.
    Affinity labelling studies with ^{109}Cadmium in cadmium-induced
    testical injury in rats.
    J. Reprod. Fert. 38, 293, 1974.

55. CHEN R.W. and GANTHER H.E.
    Some properties of a unique cadmium-binding moiety in the soluble
    fraction of rat testes.
    Environ. Physiol. Biochem. 5, 235, 1975.

56. CHERIAN M.G. and VOSTAL J.J.
    On the role of cadmium-binding protein in the transport and excretion
    of cadmium in the rat.
    Tox. Appl. Pharmacol. 29, 141, 1974.

57. CHERIAN M.G. and SHAIKH Z.A.
    Metabolism of intravenously injected cadmium-binding protein.
    Biochem. Biophys. Res. Comm. 65, 863, 1975.

58. CHERIAN M.G. and GOYER R.A.
    Toxic effects of cadmium - metallothionein in rats.
    Abstract N° 155, p. 127 in Abstracts of papers.
    Society of Toxicology. Fifteenth Annual Meeting, Atlanta,
    Georgia 1976.

59. CHERNOFF N.
    The teratogenic effects of cadmium in the rat.
    Toxicol. Appl. Pharmacol. 22, 313, 1972.

60. CHERNOFF N.
    Teratogenic effects of cadmium in rats.
    Teratology 8, 29, 1973.

61. CHIAPPINO G. and BARONI M.
     Morphological signs of hyperactivity of the renin-aldosterone
     system in cadmium-induced experimental hypertension.
     Med. Lav. (in Italian), 60, 297, 1969.

62. CHIAPPINO G. and PERNIS B.
     Effetti del cadmio sur catabolismo renale delle proteine.
     Med. d. Lavoro, 61, 424, 1970.

63. CHIQUOINE A.D.
     Observations on the early events of cadmium necrosis of the testis.
     Anat. Rec. 149, 23, 1964.

64. CHIQUOINE A.D.
     The effect of cadmium chloride on the pregnant albino mouse.
     J. Reprod. Fert. 10, 263, 1965.

65. CHIZHIKIV D.M.
     Cadmium.
     Pergamon Press, N.Y. 1966.

66. CHOWDHURY P. and LOURIA D.B.
     Influence of cadmium and other trace metals on human α_1- antitrypsin:
     an in vitro study.
     Science, 191, 480, 1976.

67. CHRISTENSEN H.E., LUGINBYHL T.T. , Editor
     The toxic substances List. 1974 Edition.
     U.S. Department of Health Education and Welfare.

68. CIKRT M. and TICHY M.
     Excretion of cadmium through bile and intestinal wall in rats.
     Brit. J. Industr. Med. 31, 134, 1974.

69. CLARKSON T.W. and KENCH J.E.
     Urinary excretion of amino acids by men absorbing heavy metals.
     Biochem. J. 62, 361, 1956.

70. COLUCCI A.V., WINGE D. and KRASNO J.
     Cadmium accumulation in rat liver.
     Arch. Environ. Health 30, 153, 1975.

142

71. COMMISSION DES COMMUNAUTES EUROPEENNES
    Le mercure, le cadmium et le chrome aux Pays-Bas.
    Environnement et Qualité de la Vie, N° 2, Septembre 1973.

72. COMMISSION DES COMMUNAUTES EUROPEENNES
    Proposition de directive du Conseil relative à la qualité des eaux
    destinées à la consommation humaine.
    J. Officiel des Communautés Européennes, 18, 2, 1975.

73. COOK J.A., HOFFMANN E.O. and DILUZIO N.R.
    Influence of lead and cadmium on the susceptibility of rats to
    bacterial challenge.
    Proc. Soc. Exp. Biol. Med.150, 741, 1975.

74. COTZIAS. G.C., BORG D.C. and SELLECK B.
    Virtual absence of turnover in cadmium metabolism: Cd^{109} studies
    in the mouse.
    Amer. J. Physiol. 201, 927, 1961.

75. CRAWFORD M.D., GARDNER M.J. and MORRIS J.N.
    Mortality and hardness of local water supplies.
    The Lancet i, 827, 1968.

76. CREASON J.P., McNULTY O., HEIDERSHEIT L.T., SWANSON D.H. and BUECHLEY RW.
    Roadside gradients in atmospheric concentrations of cadmium, lead
    and zinc.
    in Trace Substances in Environmental Health, V.A. Symposium,
    Hemphill D.D. Ed. Univ. of Missouri Press, Columbia 1972.

77. CURRY A.S. and KNOTT A.R.
    "Normal" levels of cadmium in human liver and kidney in England.
    Clin. Chim. Acta 30, 115, 1970.

78. DALHAMM I. and FRIBERG L.
    The effects of cadmium on blood pressure and respiration and the
    use of dimercaprol (BAL) as antidote.
    Acta Pharmacol. 10, 199, 1954.

79. DAVIES J.M.
    Mortality among workers at two copper works where cadmium was in use.
    Brit. J. Prevent. Soc. Med. 26, 59, 1972.

80. DAWSON E.B., CROFT H.A., CLARK R.R. and MC GANITY W.J.
    Study of seasonal variation in nine cations of normal term placentas.
    Amer. J. Obstet and Gynec. 102, 354, 1968.

81. DECKER C.F., BYERRUM R.V. and HOPPERT C.A.
    A study of the distribution and retention of cadmium -115 in the
    albino rat.
    Arch. Biochem. Biophys. 66, 140, 1957.

82. DECKER L.E., BYERRUM R.E., DECKER C.F., HOPPERT C.A. and LANGHAM R.F.
    Chronic toxicity studies
    I. Cadmium administered in drinking water to rats.
    A.M.A. Arch. Ind. Health 18, 228, 1958.

83. DEKNUDT G.H. and LEONARD A.
    Cytogenetic investigations on leucocytes of workers from a cadmium
    plant.
    Environ. Physiol. Biochem. 5, 319; 1975.

84. DELVES H.T., EPHERD G. and VINTER P.
    Determination of eleven metals in small samples of blood by sequen-
    tial solvent extraction and atomic-absorption spectrophotometry.
    Analyst, 96, 260, 1971.

85. DELVES H., BICKNELL J. and CLAYTON B.
    The excessive ingestion of lead and other metals by children.
    in Proceedings Int. Symp. Environ. Health Aspect Lead.
    Commission of the European Communities, Luxembourg 1973.

86. DEMUYNCK M., RAHN K.A., JANSSENS M. and DAMS R.
    Chemical analyses of airborne particulate matter during a period
    of unusually high pollution.
    Atmospheric Environment 10, 21, 1976.

87. DENCKER L.
    Possible mechanisms of cadmium fetotoxicity in golden hamsters
    and mice: uptake by the embryo, placenta and ovary.
    J. Reprod. Fert. 44, 461, 1975.

88. DERIVAUX J.

Nécrose testiculaire par le chlorure de cadmium chez le mouton.
(Personal communication 1973).

89. DIAMOND E.M. and KENCH J.E.

Effects of cadmium on the respiration of rat liver mitochondria.
Environ. Physiol. Biochem. 4, 280, 1974.

90. DOOLAN K.J. and SMYTHE L.E.

Cadmium content of some new south wales waters.
Search 4, 162, 1973.

91. DORN C.R., PIERCE J.O., CHASE G.R., and PHILLIPS P.E.

Environmental contamination by lead, cadmium, zinc and copper in
a new lead-producing area.
Environ. Res. 9, 159, 1975.

92. DREIZEN S., LERY B.M., NIEDERMEIER W. and GRIGGS J.H.

Comparative concentrations of selected trace metals in human and
marmoset saliva.
Arch. Oral Biol. 15, 179, 1970.

93. DUGGAN R.E. and CORNELIUSSEN P.E.

Dietary intake of pesticide chemicals in the United States.
(III) June 1968 - April 1970
Pestic. Monit. J. 5, 331, 1972.

94. DUNPHY B.

Acute occupational cadmium exposure, a critical review of the
literature.
J. Occup. Med. 9, 22, 1967.

95. DURUM W.H., HEM J.H. and HEIDEL S.G.

Reconnaissance of selected minor elements in surface waters of the
United States; Geological Survey Circular 643, U.S. Department of the
Interior, Washington D.C., October 1970, cited by Friberg et al (129)

96. EDIGER R.D. and COLEMAN R.L.

Determination of cadmium in blood by a delves cup technique.
Atomic Absorpt. Newsl. 12, 3, 1973.

97. EINBRODT H.J., ROSMANITH J. and PRAJSNAR D.
    Der Cadmiumgehalt im Blut und Reuchgewohnheiten.
    Naturwissenschaften 63, 148, 1976.

98. ENGBERG Å. and BRO-RASMUSSEN F.
    Sources of direct cadmium contamination in food.
    in European Colloquium: Problems of the contamination of man and
    his environment by mercury and cadmium. Luxembourg, July 1973.

99. ENGBERG Å. and BRO-RASMUSSEN F.
    Study on information already available in the literature on food con-
    tamination caused by lead and cadmium in ceramic household containers
    CEC Doc. N° V/F/3799/74 e, Luxembourg, 1975.

100. ENVIRONMENTAL PROTECTION AGENCY OFFICE OF RESEARCH AND DEVELOPMENT
     Scientific and technical assessment report on cadmium.
     Washington D.C., 1974.

101. EPSTEIN S.S., ARNOLD E., ANDREAS J., BASS W. and BISHOP Y.
     Detection of chemical mutagens by the dominant lethal assay in
     the mouse.
     Tox. Appl. Pharmacol. 23, 288, 1972.

102. ESCHNAUER H.
     Bestimmung von Cadmium im Wein.
     Z. Lebensm. Unters-Forsch. 127, 4, 1965.

103. ESSING H.B., SCHALLER K.H., SZADKOWSKI D. and LEHNERT G.
     Usuelle Cadmiumbelastung durch Nahrungsmittel und Getränke
     Arch. Hyg. Bakt. 153, 490, 1969.

104. EVANS G.W., MAJORS P.F. and CORNATZER W.E.
     Mechanism for cadmium and zinc antagonism of copper metabolism.
     Biochem. Biophys. Res. Comm. 40, 1142, 1970.

105. EXON J.H., PATTON N.M. and KOLLER L.D.
     Hexamitiasis in cadmium-exposed mice.
     Arch. Environ. Health. 31, 463, 1975.

106. EYBEL V. and SYKORA J.
     Die Schutzwirkung von Chelatblidern bei der akuten Kadmiumcloridver-
     giftung. Acta Biol. med. germ. 16, 61, 1966.

107. FAIRHALL L.T.

Industrial toxicology.

2nd ed. Williams and Wilkins, Baltimore, Md, 1957.

108. FALCHUK K.H., EVENSON M. and VALLEE B.L.

A multichannel atomic absorption instrument: simultaneous analysis of zinc, copper and cadmium in biologic materials.

Anal. Biochem. 62, 255, 1974.

109. FAVINO A. and NAZARI G.

Renal lesions induced by a single subcutaneous $CdCl_2$ injection in rat.

Lav. Um. 19, 167, 1967.

110. FAVINO A., CANDURA F., CHIAPPINO G. and CAVALLERI A.

Study on the androgen function of men exposed to cadmium.

Med. Lav. 59, 105, 1968.

111. FERM V.H.

The synteratogenic effect of lead and cadmium.

Experientia 25, 56, 1967.

112. FERM V.H. and CARPENTER S.J.

Teratogenic effect of cadmium and its inhibition by zinc.

Nature 216, 1123, 1967.

113. FERM V.H. and CARPENTER S.J.

The relationship of cadmium and zinc in experimental teratogenesis.

Lab. Inv. 18, 429, 1968.

114. FERM V.H. and CARPENTER S.J.

The permeability of the hamster placenta to radioactive cadmium.

J. Embryol. Exp. Morphol. 22, 107, 1969.

115. FERM V.H.

Developmental malformations induced by cadmium.

Biol. Neonat. 19, 101, 1971.

116. FESTY B., Personal communication.

117. FISCHER G.M. and THIND G.S.

Tissue cadmium and water content of normal and cadmium hypertensive rabbits. Arch. Environ. Health 23, 107, 1971.

118. FITZHUGH O.G. and MEILLER F.H.
     The chronic toxicity of cadmium.
     J. Pharmacol. Exp. Ther. 72, 15, 1941.

119. FLEISHER M., SAROLIM A.F., FASSETT D.W., HAMMOND P., SCHACKLETTE H.T.,
     NISBET I.C.T. and EPSTEIN S., Environmental impact of cadmium: a review
     by the panel on hazardous trace substances.
     Environmental Health perspectives, 253, May 1974.

120. FLICK D.F., KRAYBILL H.F. and DIMITROFF J.M.
     Toxic effects of cadmium: a review.
     Environm. Res. 4, 71, 1971.

121. FOOD and DRUG ADMINISTRATION – U.S.A.
     FDA's total diet survey, 1973, 1974.

122. FOSTER C.L. and CAMERON E.
     Observations on the histological effects of sub–lethal doses of
     cadmium chloride in the rabbit.
     II. The effect on the kidney cortex.
     J. Anat. 97, 281, 1963.

123. FOX M.R.S. and FRY B.E.Jr
     Cadmium toxicity decreased by dietary ascorbic acid supplements.
     Science 169, 989, 1970.

124. FOX M.R.S., FRY B.E. Jr, HARLAND B.F., SCHERTAL M.E. and WEEKS C.E.
     Effect of ascorbic acid on cadmium toxicity in the young coturnix.
     J. Nutr. 101, 1295, 1971.

125. FOWLER B.A., JONES H.S., BROWN H.W. and HASEMAN J.K.
     The morphologic effects of chronic cadmium administration on the
     renal vasculature of rats given low and normal calcium diets.
     Tox. Appl. Pharmacol. 34, 233, 1975.

126. FRIBERG L.
     Proteinuria and kidney injury among workmen exposed to cadmium and
     nickel dust.
     J. Ind. Hyg. Toxicol. 30, 32, 1948.

127. FRIBERG L.
     Health hazards in the manufacture of alkaline accumulators with
     special reference to chronic cadmium poisoning.
     Acta med. Scand. 138, Suppl. 240, 1950.

128. FRIBERG L.

Iron and liver administration in chronic cadmium poisoning and studies on the distribution and excretion of cadmium. Experimental investigations in rabbits.
Acta Pharmacol. 11, 168, 1955.

129. FRIBERG L., PISCATOR M., NORDBERG G.F., KJELLSTRÖM T.

Cadmium in the environment 2nd Edition.
CRC Press Inc., Cleveland, OHIO 1974.

130. FRICKENHAUS B., LIPPAL J., GORDON Th. and EINBRODT H.J.

Blutdruck und Pulsfrequenz bei oraler Belastung mit Cadmiumsulfid im Tierversuch.
Zbl. Bakt. Hyg. I Abt. Orig. B. 161, 371, 1976.

131. FUKUSHIMA M.

Environmental pollution by cadmium and its health effects: an epidemiological approach to the "Itai-Itai" disease.
in New Methods in Environmental Chemistry and Toxicology
(F. Coulston et al. Eds). Internation Academic Printing Co, Totsuka, Tokyo, 1973.

132. FUKUSHIMA M.

Cadmium content in various foodstuffs.
in Kankyo Hoken Report N° 11 Japanese Association of Public Health April 1972 cited by Friberg et al (129).

133. FUKUYAMA Y. and KUBOTA K.

Relationship between proteinuria and heavy metal pollution in the Itai-Itai disease district.
Med. Biol. 85, 103, 1972. Cited by Friberg et al. (129).

134. FULKERSON W. and GOELHER H.E.

Cadmium, the dissipated element.
Oak Ridge National Laboratory 1973.

135. FULKERSON W.

Cadmium - The dissipated element - revisited.
Oak Ridge National Laboratory, Tennessee, 1975.

136. GABBIANI G.

Action of cadmium chloride on sensory ganglia.
Experientia, 22, 261, 1966.

137. GABBIANI G., BAIC D. and DEZIEL C.

Studies on tolerance and ionic antagonism for cadmium or mercury.

Can. J. Physiol. Pharmacol. 45, 443, 1967.

138. GABBIANI G., BAIC D. and DEZREL C.

Toxicity of cadmium for the central nervous system.

Exp. Neurol. 18, 154, 1967.

139. GABBIANI G., GREGORY A. and BAIC D.

Cadmium-induced selective lesions of sensory ganglia.

J. Neuropathol. Exp. Neurol. 26, 498, 1967.

140. GALE T.F.

The interaction of mercury with cadmium and zinc in mammalian embryonic development.

Environ. Res. 6, 95, 1973.

141. GERVAIS J. and DELPECH P.

L'intoxication cadmique.

Arch. Mal. Prof. 24, 803, 1963.

142. GHAFGHAZI T. and MENNEAR J.H.

Effects of acute and subacute cadmium administration on carbohydrate metabolism in mice.

Tox. Appl. Pharmacol. 26. 231, 1973.

143. GILLIAVOD N. and LEONARD A.

Mutagenicity tests with cadmium in the mouse.

Toxicology 5, 43, 1975.

144. GINN J.T. and VOLKER J.F.

Effect of cadmium and fluorine on the rat dentition.

Proc. Soc. Exp. Biol. Med. 59, 189, 1944.

145. GIROD C.

A propos de l'influence du chlorure de cadmium sur le testicule: recherches chez le singe macacus irus. F.Cuv.

C.R. Séances Soc. Biol. Fil. 158, 297, 1964.

146. GIROD C.

Etude des cellules gonadotropes antéhypophysaires du singe macacus: irus F. Cuv., après administration de chlorure de cadmium.

C.R. Séances Soc. Biol. Fil. 158, 948, 1964.

147. GLAUSER S.C., BELLO C.T., and GLAUSER E.M.
Blood-cadmium levels in normotensive and untreated hypertensive humans.
The Lancet 1, 717, 1976.

148. GLEASON M.N., GOSSELIN R.S., HODGE H.C. and SMITH R.P.
Clinical toxicology of commercial products: acute poisoning.
The Williams and Wilkins Company, Baltimore, 1969.

149. GOODMAN G. and ROBERTS T.M.
Plants and soils as indicators of metals in the air.
Nature, 231, 287, 1971.

150. GOYER R.A., TSUCHIYA K., LEONARD D.L. and KAHYO H.
Aminoaciduria in Japanese workers in the lead and cadmium industries.
Am. J. Clin. Pathol. 57, 635, 1972.

151. GUNN S.A., GOULD T.C. and ANDERSON W.A.D.
Zinc protection against cadmium injury to rat testis.
Arch. Pathol. 71, 274, 1961.

152. GUNN S.A., GOULD T.C. and ANDERSON W.A.D.
The selective injurious response of testicular and epididymal blood vessels to cadmium and its prevention by zinc.
Am. J. Path. 42, 685, 1963.

153. GUNN S.A., GOULD T.C. and ANDERSON W.A.D.
Effect of zinc on cancerogenesis by cadmium.
Proc. Soc. Exp. Biol. Med. 115, 653, 1964.

154. GUNN S.A. GOULD T.C. and ANDERSON W.A.D.
Specific response of mesenchymal tissue to cancerogenesis by cadmium.
Arch. Pathol. 83, 493, 1967.

155. GUNN SA.A, GOULD T.C. and ANDERSON W.A.D.
Selectivity of organ response to cadmium injury and various protective measures.
J. Pathol. Bacteriol. 96, 89, 1968.

156. GUNN S.A., GOULD T.C. and ANDERSON W.A.D.
Mechanisms of zinc, cysteine and selenium protection against cadmium induced vascular injury to mouse testis.
J. Reprod. Fert. 15, 65, 1968.

157. GUNN S.A. and GOULD T.C.
    Cadmium and other mineral elements in the testis, Vol. 3.
    JOHNSON A.D., GOMES W.R. and VANDEMARK N.L. Eds,
    Academic Press, New York 1970.

158. GUTHRIE B.E.
    Daily dietary intakes of zinc, copper, manganese, chromium and
    cadmium by some New Zealand women.
    Proc. Univ. Otago Med. School., 51, 47, 1973.

159. GUTHRIE B.E.
    Chromium, manganese, copper, zinc and cadmium content of New
    Zealand foods.
    New Zealand Med. J., 82, 418, 1975.

160. HADDOW A., ROE F.J.C., DUKES C.E. and MITCHLEY B.C.V.
    Cadmium neoplasia — sarcomata at the site of injection of cadmium
    sulphate in rats and mice.
    Brit. J. Cancer 18, 667, 1964.

161. HADLEY W.M. and MIYA T.S.
    Cadmium potentiated hexobarbital sleep time in male and female
    albino rats and mice.
    Toxicol. Appl. Pharmacol. 22, 311, 1972.

162. HAMILTON D.L. and VALBERG L.S.
    Relationship between cadmium and iron absorption.
    Amer. J. Physiol. 227, 1033, 1974.

163. HAMMER D.I., FINKLEA J.F., HENDRICKX R.H., SHY C.M. and HORTON R.J.M.
    Hair trace metal levels and environmental exposure.
    Amer. J. Epidemiol. 93, 84, 1971.

164. HAMMER D.I., CALOCCI A.V., HASSELBLAD V., WILLIAMS M.E. and PINKERSON C.
    Cadmium and lead in autopsy tissue.
    J. Occupat. Med. 15, 956, 1973.

165. HANLON D.P. and URBAN J.
    The permeability of the hamster placenta to radioactive cadmium.
    J. Embryol. Exp. Morphol. 22, 107, 1969.

166. HARDY H.L. and SKINNER J.B.
    The possibility of chronic cadmium poisoning.
    J. Ind. Hyg. Toxicol. 29, 321, 1947.

167. HARRISON H.E., BUNTING H., ORDWAY N. and ALBRINK W.S.
    The effects and treatment of inhalation of cadmium chloride in the
    dog.
    J. Ind. Hyg. Toxicol. 29, 302, 1947.

168. HARRISON P.R. and WINCHESTER J.W.
    Area-wide distribution of lead, copper and cadmium in air particu-
    lates from Chicago and Northwest Indiana.
    Atmos. Environ. 5, 863, 1971.

169. HARVEY T.C., THOMAS B.J., McLELLAN J.S. and FREMLIN J.H.
    Measurement of liver-cadmium concentrations in patients and indus-
    trial workers by neutron-activation analysis.
    The Lancet, June, 1269, 1975.

170. HAVRDOVA J., CIKRT M. and TICHY M.
    Binding of cadmium and mercury in the rat bile: studies using
    gel filtration.
    Acta Pharmacol. et toxicol. 34, 246, 1974.

171. HAVRE G.N., UNDERDAL B., and CHRISTIANSEN C.
    The content of lead and some other heavy elements in different fish
    species from a fjord in Western Norway.
    Int. Symp. Environ. Health Aspect of Lead.
    Commission of the European Communities, Luxembourg, 1973.

172. HAYES J.A., SNIDER G.L. and PALMER K.C.
    The evolution of biochemical damage in the rat lung after acute
    cadmium exposure.
    Amer. Rev. Resp. Dis. 113, 121, 1976.

173. HEATH J.C., DANCEL M.R., DINGLE J.T. and WEBB. M.
    Cadmium as a carcinogen.
    Nature, 193, 592, 1962.

174. HEINDRYCKX R., DEMUYNCK M., DAMS R., JANSSENS M. and RAHN R.A.
     Mercury and cadmium in Belgian aerosols.
     in European Colloquium – Problems of the contamination of man and
     his environment by mercury and Cadmium.
     CEC – Luxembourg, 1973.

175. HENKE G., SACHS H.W. and BOHN G.
     Cadmiumbestimmungen in Leber und Nieren von Kindern und Jugendlichen
     durch Neutronaktievierunganalyse.
     Arch. Toxikol. 26, 8, 1970

176. HEULLY F., LAMY P., PERNOT C. and COUILLAULT S.
     Intoxications aiguës par vapeurs de cadmium.
     Arch. Mal. Prof. 24, 547, 1963.

177. HICKEY R.J., SCHOFF E.P. and CLELLAND R.C.
     Relationship between air pollution and certain chronic disease
     death rates.
     Arch. Environ. Health, 15, 728, 1967.

178. HICKS D.J., MIYA T.S. and SCHNELL R.C.
     Sex-related differences in cadmium toxicity in rats.
     Abstract N° 154, p. 127 in Abstracts of papers. Society of
     Toxicology. Fifteenth Annual Meeting, Atlanta, Georgia 1976.

179. HINE C.H., WRIGHT J. and GOODMAN D.
     Tissue levels of cadmium in different disease states.
     Abstract N° 100. Twelfth Annual Meeting the Society of Toxicology
     New York 1973.

180. HIRST R.N.Jr., PERRY H.M.Jr., CRUZ M.G. and PIERCE J.A.
     Elevated cadmium concentration in emphysematous lungs.
     Amer. Rev. Resp. Disease 108, 30, 1973.

181. HISASHI Y.
     Cadmium poisoning, susceptibility in different strains of mice and
     effect on pregnancy.
     Sch. Med. Nihon Univ. Tokyo, Japan, Nichidai Igaku Zasshi 31,
     1201, 1972.

182. HISASHI Y. and LIDASS.

Difference in the detrimental effects of cadmium and arsenic on chick embryo femurs.

Jap. J. Pharmacol. 23, Suppl. 130, 1973.

183. HISE E.C. and FULKERSON W.

Chapter VI Environmental Impact of cadmium flow. in Cadmium, The dissipated element, Fulkerson W. and Goelher H.E., Editors Oak Ridge National Library 1973.

184. HODGEN G.D., GOMES W.R. and VANDEMARK N.L.

Carbonic anhydrase isoenzymes in rat erythrocytes, kidney and testis.

Fed. Proc. 28, 773, 1969.

185. HOFFMANN E.O., COOK J.A., DILUZIO N.R. and COOVER J.A.

The effects of acute cadmium administration in the liver and kidney of the rat. Light and electron microscopic studies.

Labor. Invest. 32, 655, 1975.

186. HOLDEN H.

Cadmium toxicology.

Lancet 2, 57, 1969.

187. HOLMBERG R.E. and FERM V.H.

Interrelationships of selenium, cadmium and arsenic in mammalian teratogenesis.

Arch. Environ. Health. 18, 873, 1969.

188. HOSTE J., DAMS R., BLOCK C., DEMUYNCK H., HEINDRYCKS R.

Study of National air pollution by combustion.

Part. I: Inorganic composition of airborne particulate matter. Instituut voor nucleaire Wetenschappen.

Rijks Universiteit Gent, Belgium, 1974.

189. HUMPERDINCK K.

Kadmium und Lungenkrebs.

Med. Klin. 63, 948, 1968.

190. HUNT W.F.Jr., PINKERTON C., McNULTY O. and CREASON J.
     A study of trace element pollution of air in 77 midwestern cities.
     In Trace Substances in Environmental Health, vol. 4, Hemphill
     D.E., Ed. University of Missouri press, Columbia 1971.

191. HYGIENIC GUIDE SERIES: CADMIUM
     American Industrial Hygiene Association. September 1944.

192. IARC Working groups on the Evaluation of the carcinogenic risk of
     chemicals to man.
     IARC Monograph, vol. 2, Lyon 1973.

193. IMBUS H.R., CHOLAK J., MILLER L.H. and STERLING T.
     Boron, cadmium, chromium and nickel in blood and urine.
     Arch. Environ. Health. 6, 286, 1963.

194. ISHIZAKI A., FUKUSHIMA M. and SAKAMOTO M.
     Distribution of cadmium in biological materials.
     II. Cadmium and zinc contents of foodstuffs.
     Jap. J. Hyg. 25, 207, 1970 cited by Friberg et al. (129)

195. ISHIZAKI A., FUKUSHIMA M. and SAKAMOTO M.
     Contents of cadmium and zinc in organs of patients with Itai-Itai
     disease residents in Hokuriku district (in Japanese).
     Jap. J. Hyg. 26, 268, 1971.

196. ISHIZU SUMIKO, MINAMI MASAYASU, SUZUKI AKIO, YAMADA MINEKO, YMAMURA KOTARO
     Teratogenic effect of cadmium.
     Dep. Public. Health, Tokyo Women's Med. Coll, Tokyo, Japan
     Ind. Health 11, 127, 1973.

197. ITO T. and SAWAUCHI K.
     Inhibitory effects on cadmium-induced testicular damage by pre-
     treatment with smaller cadmium dose.
     Okajimas Folia Anat. Jap. 42, 107, 196, cited by Friberg et al. (129)

198. ITOKAWA Y., ABE T. and TANAKA S.
     Bone changes in experimental chronic cadmium poisoning.
     Arch. Environ. Health 26, 241, 1973.

199. ITOKAWA Y., ABE T., TABEI R. and TANAKA S.
Renal and skeletal lesions in experimental cadmium poisoning.
Arch. Environ. Health 28, 149, 1974.

200. JAAKKOLA T., TAKAHASHI H. and MIETTINEN J.
Cadmium content in sea water, bottom sediments, fish, lichen and
elk in Finland.
Page 230 in Environmental quality and safety.
Global aspects of chemistry, toxicology and technology as applied
to the environment, Vol. II
F. Coulston and F. Korte, Editors. Academic Press, New York 1973.

201. JOHNSON M.H.
The effect of cadmium chloride on the blood-testis barrier of the
guinea-pig.
U. Reprod. Fert. 19, 551, 1969.

202. JOHNSON A.D. and WALKER G.P.
Early actions of cadmium in the rat and domestic fowl testis.
V. Inhibition of carbonic anhydrase.
J. Reprod. Fert. 23, 463, 1970.

203. JOHNSON A.D., and SIGMAN M.B.
Early actions of cadmium in the rat and domestic fowl testis:
4. Autoradiographic location of cadmium 115 m.
J. Reprod. Fert. 24, 115, 1971.

204. JOHNSON C.A.
The determination of some toxic metals in human liver as a guide
to normal levels in New Zealand.
Part. I. Determination of Bi, Cd, Cr, Co, Cu, Pb, Mn, Ni, Ag,
Tl and Zn.
Analyt. Chim. Acta. 81, 69, 1976.

205. JOHNSTON R.E., MIYA T.S. and SCHNELL R.C.
Cadmium potentiation of drug response-role of the liver.
Biochem. Pharmacol. 24, 877, 1975.

206. JUST J. and KELUS J.
Cadmium in the air atmosphere of ten selected cities in Poland.
Rocz. Panstw. Zakl. Hig. 22, 249, 1971.

207. KAGI J.H.R., VALLEE B.L.

Metallothionein: a cadmium and zinc-containing protein from equine renal cortex.

J. Biol. Chem. 236, 2435, 1961.

208. KAJIKAWA K. et al.

A pathological study of Itai-Itai disease.

J. Juzen Med. Soc. cited by Nordberg 1975.

209. KANISAWA M. and SCHROEDER H.A.

Life term studies on the effect of trace elements on spontaneous tumours in mice and rats.

Cancer res. 29, 892, 1969.

210. KANISAWA M. and SCHROEDER H.A.

Renal arteriolar changes in hypertensive rats given cadmium in drinking water.

Exp. Mol. Pathol. 10, 81, 1969.

211. KAR A.B., DAS R.P. and MUKERJI B.

Prevention of cadmium induced changes in the gonads of rat by zinc and selenium. A study in antagonism between metals in the biological ststem. Proc. Natl. Inst. Sci. India. Part B Biol. ScI 26, suppl. 40, 1960.

212. KAR A.B. and DAS R.P.

The nature of protective action of selenium on cadmium-induced degeneration of rat testis.

Proc. Natn. Inst. Sci. India, B, 29, 297, 1963.

213. KARHAUSEN L.R.

L'absorption intestinale du cadmium et du mercure , in European Colloquium: Problems of contamination of man and his environment by mercury and cadmium.

Luxembourg, 1973.

214. KARLICEK V. and TOPOLCAN O.

Cadmium in kidneys of patients with essential and renal hypertension.

Cas Lek. Ces. 113, 41, 1973.

(abstract in Excerpta Medica sect XVIII, 21, 1974).

215. KATO T. and KAWANO S.
Review of past and present of Itai-Itai disease. On the process
of research development.
Curr. Med. 16, 29, 1968 (cited by Friberg et al. 129)

216. KAWAI K.
Pathological findings in experimental cadmium intoxication.
Proceeding of cadmium symposia pp. 52-53 Jap. Assoc. Publ. Health
Tokyo 1973.

217. KAZANTIZIS G.
Respiratory function in men casting cadmium alloys.
I. Assessment of ventilatory function.
Brit. J. Ind. Med. 13, 30, 1956.

218. KAZANTZIS G.
Induction of sarcoma in the rat by cadmium sulphide pigment.
Nature, 198, 1213, 1963.

219. KAZANTIZIS G., FLYNN F.V., SPOWAGE J.S. and TROTT D.G.
Renal tubular malfunction and pulmonary emphysema in cadmium
pigment workers.
Q. J. Med. 32, 165, 1963.

220. KAZNATIZIS G. and HANBURY W.J.
The induction of sarcoma in the rat by cadmium sulphide and by
cadmium oxide.
Brit. J. Cancer 21, 190, 1966.

221. KEINO H.
Effect of $CdSO_4$ on closure of the neutral tube in frogs.
Teratology 8, 96, 1973.

222. KEMPF Th.
Quecksilber- und Cadmiumgehalte im Wasserkreislauf.
in European Colloquium: Problems of the contamination of man
and his environment by mercury and cadmium.
Luxembourg, July 1973.

223. KENCH J.E., GAIN A.C. and SUTHERLAND E.M.
A biochemical study of the minialbumin to be found in the urine
of men and animals poisoned by cadmium.
S. Afr. Med. J. 39, 1191, 1965.

224. KENCH J.E. and SUTHERLAND E.M.
The nature and origin of the minialbumin found in cadmium-poisoned
animals.
S. Afr. Med. J. 40, 1109, 1966.

225. KIMURA M., OTAKI N., YOSHIKI S., SUZUKI M., HORIUCHI N. and SUDA T.
The isolation of metallothionein and its protective role in cad-
mium poisoning.
Arch. Biochem. Biophys. 165, 340, 1974.

226. KING E.
An environmental study of casting copper-cadmium alloys.
Brit. J. Ind. Med. 12, 198, 1955.

227. KIPLING M.D. and WATERHOUSE J.A.H.
Cadmium and prostatic carcinoma.
Lancet 1, 730, 1967.

228. KJELLSTRÖM T., LIND B., LINNMAN L. and NORDBERG G.
A comparative study on methods for cadmium analysis of grain with
an application to pollution evaluation.
Environ. Res. 8, 92, 1974.

229. KJELLSTRÖM T., LIND B., LINNMAN L., and ELINDER C.G.
Variation of cadmium concentration in Swedish wheat and barley.
An indicator of changes in daily cadmium intake during the 20th
century.
Arch. Environ. Health 31, 321, 1975.

230. KJELLSTRÖM T., EVRIN P.E. and RAHNSTER B.
Dose-response analysis of cadmium-induced tubular proteinuria (1)
A study of workers exposed to cadmium in a swedish battery factory.
Env. Res. (in Press).

231. KJELLSTRÖM T., SHIROISHI K. and EVRIN P.E.

Urinary β_2 micro-globulin excretion among people exposed to cadmium in the general environment.

An epidemiological study in cooperation between Japan and Sweden.

Env. Res. (in Press).

232. KLOKE A.

Cadmium in Boden und Pflanze.

Ein Beitrag sum Thema "Umweltschutz".

Nachrichtenbl. Deutsch. Pflanzenschutzd. 22, 164, 1971.

233. KNEIP T.J., EISENBUD N., STREHLOW C.D. and FREUDENTHAL P.C.

Airborne particulates in New York City.

J. Air Pollut. Contr. Ass. 20, 144, 1970.

234. KNORRE D.

Enzymatische Untersuchungen an Rattenhoden innerhald 30 Tagen nach Cadmium induzierte Nekrose.

Virchows Arch. Abt. A. Path. Anat. 345, 255, 1968.

235. KNORRE, von D.

Zur Induktion von Hautsarkomen bei der Albinoratte durch Kadmiumchlorid.

Arch. Gescwulstforsch. 36, 119, 1970a.

236. KNORRE, von D.

Ortliche Hautschädigungen an der Albinoratte in der Latenzperiode der Sarkomentwicklung nach Cadmiumchlorid Injektion.

Zbl. Allg. Path. 113, 192, 1970b.

237. KNORRE, von D.

Zur Induktion von Hodenzwischenzlltumoren an der Albinoratte durch Cadmiumchlorid.

Arch. Geschwulstforsch. 38, 257, 1971.

238. KOBAYASHI J., NAKAHARA D. and HASEGAWA Y.

Accumulation of cadmium in organs of mice fed on cadmium-polluted rice.

Jap. J. Hyg. 26, 401, 1971. (English Summary only)

239. KOBAYASHI J.

Relation between the "Itai-Itai" disease and the pollution of river water by cadmium from a mine.

Fifth International Water Pollution Research Conference, San Francisco, July 1970, Pergamon Press, N.Y. 1971.

240. KOLLER L.D.

Immunosuppression produced by lead, cadmium and mercury.

Am. J. Vet. Res. 34, 1457, 1973.

241. KOLLER L.D., EXON J.H. and ROAN J.G.

Antibody suppression by cadmium.

Arch. Environ. Health. 30, 598, 1975.

242. KOLLER L.D., EXON J.H. and ROAN J.G.

Humoral antibody response in mice after single dose exposure to lead or cadmium.

Proc. Soc. Exp. Biol. Med. 151, 339, 1976.

243. KOLONEL L.N.

Association of cadmium with renal cancer.

Cancer 37, 1782, 1976.

244. KOMMISSION ZUR PRÜFUNG GESUNDHEITSSCHADLICHER ARBEITSSTOFFE DER DEUTSCHEN FORSCHUNGSGEMEINSCHAFT, Gesundheitsschädliche Arbeitsstoffe, Toxikologisch arbeitsmedizinische Begründungen von MAK-Werten. Verlag Chemie 1974.

245. KROPF R. and GELDMACHER - V. MALLINCKRODT M.

Der Cadmiumgehalt von Nahrungsmitteln und die tägliche Cadmiumaufnahme.

Arch. Hyg. Bakt. 152, 218, 1968.

246. KUBOTA J., LAZAR V.A. and LOSEE F.

Copper , zinc, cadmium and lead in human blood from 19 locations in the United States.

Arch. Environ. Health 16, 788, 1968.

247. LAAMANEN A.

Functions, progress and prospects for an environmental subarctic base level station.

Work Environ. Health 9, 17, 1972.

248. LAGERWEFF J.V.

Uptake of cadmium, lead and zinc by radish from soil and air".

Soil Sci. 3, 129, 1971.

249. LAMBOT F.

Etude de l'intoxication par le cadmium de l'anguille (Anguilla anguilla) adaptée à l'eau de mer.

CIPS - Technical report 1975.

250. LANDESGEWERBEANSTALT BAYERN , 1975.

Aufstellung einer Bilanz über den Verbrauch und Verbleib von Cadmium und seiner Verbindungen in der Bundesrepublik Deutschland unter besonderer Berücksichtingung der bei Verarbeitungsprozessen und der Vernichtung bzw. Aufarbeitung von Cadmiumhaltigen Produkten Auftretender Emissionen von Cadmium und Cadmiumverbindungen und der damit verbundenen Gefahren des Überganges von Cadmium in den menschlichen Organismus.

251. LARSSON S.E. and PISCATOR M.

Effect of cadmium on skeletal tissue in normal and cadmium-deficient rats.

Isr. J. Med. Sci. 7, 495, 1971.

252. LAUWERYS R.R., BUCHET J.P. and ROELS H.A.

Comparative study of effect of inorganic lead and cadmium on blood -aminolevulinate dehydratase in man.

Brit. J. Indust. Med. 30, 359, 1973.

253. LAUWERYS R.R., BUCHET J.P. , ROELS H.A., BROUWERS J. and STANESCU D.

Epidemiological survey of workers exposed to cadmium. Effect on lung, kidney and several biological indices.

Arch. Environ. Health 28, 145, 1974.

254. LAUWERYS R., BUCHET J.P. and ROELS H.
Effets subcliniques de l'exposition humaine au cadmium.
in Proceedings International Symposium. Problems of the contamination of man and his environment by mercury and cadmium.
Commission of the European Communities, Luxembourg 1974.

255. LAUWERYS R., BUCHET J.P., ROELS H., BERLIN A. and SMEETS J.
Intercomparison program of lead, mercury and cadmium analysis in
blood, urine and aqueous solutions.
Clinical chemistry 21, 551, 1975.

256. LAUWERYS R.R., BUCHET J.P. and ROELS H.A.
The relationship between cadmium exposure or body burden and the
concentration of cadmium in blood and urine in man.
Int. Arch. Occup. Environ. Health. 36, 275, 1976.

257. LEE R.E., PATTERSON R.K. and WAGMAN J.
Particle size distribution of metal components in urban air.
Environ. Sci. Technol. 2, 288, 1968.

258. LEE R.E. Jr, GORANSON S.S., ENVIONE P.E. and MORGAN G.B.
National air surveillance cascade impactor Network II, Size distri-
bution measurements of trace metal components.
Environ. Sci. Technol. 6, 1025, 1972.

259. LEE I.P. and DIXON R.L.
Effects of cadmium on spermatogenesis studied by volocity sedimen-
tation cell separation and seriol mating.
J. Pharmacol. Expt. Therap. 187, 641, 1973.

260. LEICESTER H.M.
The effect of cadmium on the production of caries in the rat.
J. Dent. Res. 25, 337, 1946.

261. LEHNERT G., KLARIS G., SCHALLER K.H. and HAAS T.
Cadmium determination in urine by atomic absorption spectrometry
as a screening test in industrial medicine.
Brit. J. Industr. Med. 26, 156, 1969.

262. LEMEN R.A., LEE J.S. and WAGONER J.K.
Mortality among workers exposed to cadmium.
Presented at the New York Academy of Science meeting on
Occupational Carcinogenesis, March 1975.

263. LENER J. and BIBR B.

Cadmium content in some foodstuffs in respect of its biological effects.

Vitalst. Zivilisationskr 15, 139, 1970.

264. LENER J. and MUSIL J.

Cadmium influence on the excretion of sodium by kidney.

Experientia, 26, 902, 1970.

265. LENER J. and BIBR B.

Cadmium and hypertension.

Lancet i, 970, 1971.

266. LENER J. and BIBR B.

Effects of cadmium in the sphere of experimental hypertension.

Cs. Hyg. 18, 282, 1973.

Abstract in Excepta Medica sect. XXXV, 4, 1974.

267. LEONARD A., DEKNUDT Gh. and DEPACKERE H.

Cytogenetic investigations on leucocytes of cattle intoxicated with heavy metals.

Toxicology 2, 269, 1974.

268. L'EPEE P., LAZARINI M., FRANCHOME J., N'DOKY Th. and LARRIVET C.

Contribution à l'étude de l'intoxication cadmique.

Arch. Mal. Prof. 29, 485, 1968.

269. LEVY L.S., ROE F.J.C., MALCOLM D., KAZANTZIS G., CLACK J. and PLATT. H.S.

Absence of prostatic changes in rats exposed to cadmium.

Ann. Occup. Hyg. 16, 111, 1973.

270. LEVY L.S. and CLACK J.

Further studies on the effect of cadmium on the prostate gland. I. Absence of prostatic changes in rats given oral cadmium sulphate for two years.

Ann. Occup. Hyg. 17, 205, 1975.

271. LEVY L.S., CLACK J. and ROE F.J.C.
     Further studies on the effect of cadmium on the prostate gland
     II. Absence of prostatic changes in mice given oral cadmium sul-
     phate for eighteen months.
     Ann. Occup. Hyg. 17, 213, 1975.

272. LEWIS G.P., LYLE H. and MILLER S.
     Association between elevated hepatic water-soluble protein-bound
     cadmium levels and chronic bronchitis and/or emphysema.
     Lancet ii, 1330, 1969.

273. LEWIS G.P., JUSKO W.J., COUGHLIN L.L. and HARTZ S.
     Cadmium accumulation in man: influence of smoking, occupation,
     alcoholic habit and disease.
     J. Chronic Dis. 25, 717, 1972.

274. LEWIS G.P., COUGHLIN L., JUSKO W. and HARTZ S.
     Contribution of cigarette smoking to cadmium accumulation in man.
     Lancet i, 291, 1972.

275. LINDEMAN R. and ASSENZO J.
     Correlation between water hardness and cardiovascular deaths in
     Oklahoma countries.
     Amer. J. Publ. Hlth. 54, 1071, 1964.

276. LINNMAN L., ANDERSON A., NILSSON K.O., LIND B., KJELLSTRÖM T., FRIBERG L.
     Cadmium uptake by wheat from sewages sludge used as a plant
     nutrient source.
     Arch. Environ. Health 27, 45, 1973.

277. LINNMAN L.
     Interlaboratory control study of cadmium analysis 1975.
     Personal Communication.

278. LIVINGSTON H.D.
     Measurement and distribution of zinc, cadmium and mercury in
     human kidney tissue.
     Clin. Chem. 18, 67, 1972.

279. LORKE D. und LÖSER E.

Sub-chronische orale Toxicität von Cadmium bei Ratte und Hund.

CEC-EPA-WHO International Symposium Environment and Health.

Paris, June 1974.

280. LORKE D. und LÖSER E.

Cadmiumverbindungen. Akute Toxizität.

Institut für Toxikologie. Bayer A.G., 1976.

281. LOVETT R.J., GUNTENMANN W.H., PAKKALA I.S., YOUNGS W.D., LISK D.J., BURDICK G.E. and HARRIS E.J.

A survey of the total cadmium content of 406 fishes from 49 New-York State fresh waters.

J. Fish. Res. Board. Can. 29, 1283, 1972.

282. LUCAS H.F.Jr., EDGINGTON D.N. and COLBY P.J.

Concentrations of trace elements in Great Lakes fishes.

J. Fish. Res. Board Can. 27, 677, 1970.

283. LUCIS O.J., SHAIKH Z.A. and EMBIL J.A.Jr.

Cadmium as a trace element and cadmium binding components in human cells.

Experientia 26, 1109, 1970.

284. LUCIS O.J., LUCIS R. and ATERMAN K.

The transfert of ^{109}Cd and ^{65}Zn from the mother to the newborn rat.

Fed. Proc. 30, 238, 1971.

285. LUCIS O.J., LUCIS R. and ATERMAN K.

Tumorigenesis by cadmium.

Oncology 26, 53, 1972.

286. MAC FARLAND H.N.

Inhalation toxicology.

Journal of the AOAC 58, 689, 1975.

287. MACHATA G.

Über die Normalgehalte von Cadmium, Kupfer und Zink im Blut der Wiener Bevölkerung.

Wiener Klinische Wochenschrift 87, 484, 1975.

288. MACHEMER L.

Untersuchung von Cadmiumchlorid auf Embryotoxische und Teratogene Wirkung an Ratten nach oraler Verabreichung.
Institüt für Toxikologie, Bayer A.G., 1975.

289. MAEKAWA K., TSUNERANI Y. and KUREMATSU Y.

Role of increased vascular permeability in cadmium injury of the testis.
Acta Anat. Nippon 41, 327, 1966.

290. MALCOLM D.

Potential carcinogenic effect of cadmium in animals and man.
Ann. Occup. Hyg. 15, 33, 1972.

291. MARGOSHES M., VALLEE B.L.

A cadmium protein from equine kidney cortex.
J. Am. Chem. Soc. 79, 4813, 1957.

292. MASIRONI R.

Trace elements in cardiovascular diseases in uses of activation analysis in studies of mineral element metabolism in man.
IAEA Techn. Rept. N° 122, Vienna 1970.

293. MASIRONI R., MIESCH A.T., CRAWFORD M.D. and HAMILTON E.I.

Geochemical environments, trace lements and cardiovascular diseases.
Bull. Wld. Health. Org. 47, 139, 1972.

294. MASIRONI R.

Trace elements in relation to cardiovascular diseases.
WHO, Geneva, 1974.

295. MASON K.E., BROWN J.A., YOUNG J.O. et al.

Cadmium-induced injury of the rat testis.
Anat. Rec. 149, 135, 1964.

296. MASON K.E. and YOUNG J.O.

Effectiveness of selenium and zinc in protecting against cadmium induced injury of the rat testis.
In Selenium in Biomedicine p. 383, Ed. O.H. Muth et al.
Avi Publishing Co, Westport, Conn. 1967.

297. MATERNE D., LAUWERYS R., BUCHET J.P., ROELS H., BROUWERS J. and
STANESCU D.
Investigations sur le risques résultant de l'exposition au cadmium
dans deux entreprises de production et deux entreprises d'utilisation
du cadmium.
Cahiers de Médecine du Travail, XII, 1, 1975.

298. McCABE L.J., SYMONS J.M., LEE R.D. and ROBECK G.G.
Survey of community water supply systems.
J. Amer. Water Works Assoc. 62, 670, 1970.

299. McDERMID (AAS Europe Ltd),
Review of the uses of cadmium presented at the European Zinc
Producers Meeting, Madrid, 1974.

300. McKEE J.E. and WOLF H.W.
Water quality criteria, 2nd ed., The Resources Agency of California
state Water Quality Control Board, California 1963.

301. McKENZIE J.M. and KAY D.L.
Urinary excretion of cadmium, zinc and copper in normotensive and
hypertensive women.
New Zealand Med. J. 79, 68, 1973.

302. McKENZIE J.M.
Tissue concentration of cadmium, zinc and copper from autopsy
samples.
New Zealand Med. J. 79, 1016, 1974.

303. McLELLAN J.S., THOMAS B.J., FREMLIN J.H. and HARVEY T.C.
Cadmium. Its in vivo detection in man
Phys. Med. Biol. 20, 88, 1975.

304. MEGO J.L. and CAIN J.A.
An effect of cadmium on heterolysosome formation and function
in mice.
Biochem. Pharmacol. 24, 1227, 1975.

305. MENDEN E.E., ELIA V.J., MICHAEL L.W. and PETERING H.G.
Distribution of cadmium and nickel of tobacco during cigarette
smoking.
Environ. Sci. Technol. 6, 830, 1972.

306. MERALI Z. and SINGHAL R.L.

Protective effect of selenium on certain hepatotoxic and pancrea-
toxic manifestations of subacute cadmium administration.

J. Pharmacol. Exper. Ther. 195, 58, 1975.

307. MERTZ D.P., KOSCHNICK R., WILK G.

Renale Ausscheidungsbedingungen von Cadmium beim normotensiven

und hypertensiven Menschen.

Z. Klin. Chem. Klin. Biochem. 10, 21, 1972.

308. MIETTINEN J.K.

Levels of cadmium in various foods, its environmental sources and

methods of analysis.

16th Session of the joint FAO/WHO Committee of experts on food

additive.

Geneva, April 4-12, 1972.

309. MILLER W.J., LAMPP B., POWELL G.W., SALOTTI C.A. and BLACKMON D.M.

Influence of a high level of dietary cadmium on cadmium content in

milk, excretion and cow performance.

J. Dairy Sci. 50, 1404, 1967.

310. MILLER M.L., MURTHY L., BASOM C.R. and PETERING H.G.

Alterations in hepatocytes after manipulation of the diet:

copper, zinc, and cadmium interactions.

Amer. J. Anat. 141, 23, 1974.

311. MILLER D.W., VETTER R.J., HULLINGER R.L. and SHAW S.M.

The uptake and distribution of cadmium 115 m in calcium deficient

and zinc deficient golden hamsters.

Environ. Res. 6, 473, 1973.

312. MILLER G.J., WYLIE M.J. and McKEOWN D.

Cadmium exposure and renal accumulation in an australian urban po-

pulation.

Med. J. Aust. 1, 20, 1976.

313. MINISTERIUM FÜR ARBEIT, GESUNDHEIT UND SOZIALES DES LANDES N.W.

Umweltprobleme durch Schwermetalle im Raum Stolberg.

Düsseldorf 1975a.

314. MINISTERIUM FÜR ARBEIT, GESUNDHEIT UND SOZIALES DES LANDES N.W.
Zwischenbericht über die Ermittlung der Luftverunreinigung im Raum.
Duisburg – Oberhausen – Mülheim.
Düsseldorf 1975b.

315. MOORE W.Jr., STARRA J.F., CROCKER W.C., MALANCHUK M. and ILTIS R.
Comparison of ^{115m}cadmium retention in rats following different
routes of administration.
Environ. Res. 6, 473, 1973.

316. MORGAN J.M.
Cadmium and zinc abnormalities in bronchogenic carcinoma.
Cancer 25, 1394, 1970.

317. MORGAN J.M. and BURCH H.M.
Tissue cadmium and zinc content in emphysema and bronchogenic
carcinoma.
J. Chron. Dis. 24, 107, 1971.

318. MORGAN J.M.
Normal lead and cadmium content of the human kidney.
Arch. Environ. Health 24, 364, 1972.

319. MORRIS J.N., CRAWFORD M.D. and HEADY J.A.
Hardness of local water supplies and mortality from cardiovascular
disease.
The Lancet i, 860, 1961.

320. MORTON W.E.
Hypertension and drinking water constituent in Colorado.
Amer. J. Public Health 61, 1371, 1971.

321. MULVIHILL J.E., GAMM S.H. and FERM. V.H.
Facial formation in normal and cadmium-treated golden hamsters.
J. Embryol. Exp. Morphol. 24, 393, 1971.

322. MURTHY G.K., RHEA U. and PEELER J.T.
Levels of antimony, cadmium, chromium, cobalt, manganese and zinc
in institutional total diets.
Environ. Sci. Technol. 5, 436, 1971.

323. MUSTAFA M. and CROSS C.E.

Pulmonary alveolar macrophage. Oxidative metabolism of isolated cells
and mitochondria and effect of cadmium ion on electron-and energy-
transfer reactions.

Biochem. 10, 4176, 1971.

324. NANDI M., SLONE D., JICK H., SHAPIRO S. and LEWIS G.P.

Cadmium content of cigarettes.

Lancet 2, 1329, 1969.

325. NATIONAL AIR SURVEILLANCE NETWORK

U.S. Environmental Protection Agency, 1972.

326. NATIONAL NETWORK FOR THE SURVEY OF AIR POLLUTION BY HEAVY METALS

sponsored by the belgian Ministry of Public Health,

1972-1973.

327. NATIONAL INVENTORY OF SOURCES AND EMISSIONS, CADMIUM, NICKEL AND
ASBESTOS

A report for the National Air Pollution Administration.

Contract N° CPA-22-69-131 (1968) W.E. Davis and Associates.

Leawood, Kansas Feb. 1970.

328. NICAUD P., LAFITTE A. and GROS A.

Les troubles de l'intoxication chronique par le cadmium.

Arch. Mal. Prof. 4, 192, 1942.

329. Niedersächsischen Sozialminister,

Umweltschutz in Niedersachsen.

Reinhaltung der Luft, Heft 2, Hannover, 1974.

330. NISHIZUMI M.

Electron microscopic study of cadmium nephrotoxicity in the rat.

Arch. Environ. Health 24, 215, 1972.

331. NISHIYAMA K. and NORDBERG G.

Absorption and elution of cadmium on hair.

Arch. Environ. Health 25, 92, 1972.

332. NOGAWA K. and KAWANO S.

A survey of the blood pressure of women suspected of Itai-Itai
disease.

J. Juzen Med. Soc. 77, 357, 1969 cited by Friberg et al. (129).

172

333. NOGAWA K., ISHIZAKI A., FUKUSHIMA M., SHIBATA I. and HAGINO N.
     Studies on the women with acquired Fanconi syndrome observed in the
     Ichi River basin polluted by cadmium.
     Environ. Research 10, 280, 1975.

334. NOMIYAMA K., SATO C. and YAMAMOTO A.
     Early signs of cadmium intoxication in rabbits.
     Tox. Appl. Pharmacol. 24, 625, 1973.

335. NOMIYAMA K., SUGATA Y., NOMIYAMA H. and YAMAMOTO A.
     Proceedings of the 46th annual meeting of the Japan association of
     industrial Health 1973.

336. NOMIYAMA K.
     Experimental study on cadmium intoxication.
     J. Jap. Med. Assoc. 72, 955, 1974.
     (English translation provided by the author).

337. NOMIYAMA K., SUGATA Y., MUPATA I. and NAKAGAWA S.
     Urinary low-molecular weight proteins in Itai-Itai disease.
     Environ. Res. 6, 373, 1973.

338. NOMIYAMA K.
     Toxicity of cadmium-mechanism and diagnosis.
     p. 15 in Progess in Water Technology, vol 7 (P.A. Krenkel ed.)
     Pergamon Press, Oxford, 1975.

339. NORDBERG G.F., PISCATOR M. and NORDBERG M.
     On the distribution of cadmium in blood.
     Acta Pharmacol. Toxicol. 30, 289, 1971.

340. NORDBERG G.F.
     Effects of actue and chronic cadmium exposure on the testicles
     of mice.
     Environ. Physiol. 1, 171, 1971.

341. NORDBERG G.F., PISCATOR M. and LIND B.
     Distribution of cadmium among protein fractions of mouse liver.
     Acta Pharmacol. Toxicol. 29, 456, 1971.

342. NORDBERG G.F.
     Cadmium metabolism and toxicity.
     Environm. Physiol. Biochem. 2, 7, 1972.

343. NORDBERG G.F. and NISHIYAMA K.
Whole-body and hair retention of cadmium in mice.
Arch. Environ. Health, 24, 209, 1972.

344. NORDBERG G.F. (editor)
Effects and dose/response relationships of toxic metals.
Proceeding from an International Meeting Organized by the Subcom-
mittee on the Toxicology of Metals.
Tokyo, 1974 (Elsevier under press).

345. NORDBERG G.F., GOYER R. and NORDBERG M.
Comparative toxicity of cadmium-metallothionein and cadmium
chloride on mouse kidney.
Arch. Pathol. 99, 192, 1975.

346. NORDBERG C.F.
Effects of long-term cadmium exposure on the seminal vesicles of
mice.
J. Reprod. Fert. 45, 165, 1975.

347. NORVAL E. and BUTLER L.R.P.
Trace element in the human context and their determination by atomic
absorption spectrometry.
National Physical Research Laboratory Council for Scientific and
Industrial Research.
Pretoria, November, 1974.

348. NYGAARD S.P., BONDE G.J. and HANSEN J.C.
Normalvaerdier af bly og cadmium i humant blod.
Särtryck ur Nordisk Hygienisk Tidskrift, 4, 153, 1973.

349. OCDE (ORGANISATION FOR ECONOMIC COOPERATION AND DEVELOPMENT)
Cadmium and the environment: toxicity, economy, control.
Report prepared by C.L. NOBBS. Paris 1975.

350. ODEN AV SVANTE, BERGGREN B. and ENGVALL A.G.
Heavy metals and hydrocarbons in Sludge.
Grundförbattring, 23, 55, 1970.

351. OGAWA E., SUZUKI S. and TSUZUKI H.
Radiopharmacological studies on the cadmium poisoning.
Japan J. Pharmacol. 22, 275, 1972.

352. OGAWA E., SUZUKI S., TSUZUKI H. and KAWAJIRI M.
Basal study on early diagnosis of cadmium poisoning: change in
carbonic anhydrase activity.
Japan J. Pharmacol. 23, 97, 1973.

353. OLERU U.G.
Epidemiological implications of environmental cadmium.
Amer. Indust. Hyg. Assoc. J., 229, March 1975.

354. OMAYE S.T., TAPPEL Al. L.
Effect of cadmium chloride on the rat testicular soluble seleno-
enzyme, glutathione peroxidase.
Res. Comm. Chemic. Pathol and Pharmacol. 12, 695, 1975.

355. ORGANISATION MONDIALE DE LA SANTE, série de rapports techniques N° 505
Evaluation de certains additifs alimentaires et des contaminants:
mercure, plomb et cadmium.
Genève, 1972.

356. OSTERGAARD K. and CLAUSEN P.P.
Cadmium i danske nyrer
Ugeshr. Laeg. 136, 863, 1974.

357. PAGE A.L. and BINHAM F.T.
Cadmium residues in the environment.
Residue Reviews 48, 1, 1973.

358. PARIZEK J., and ZAHOR Z.
Effect of cadmium salt on testicular tissue.
Nature, 177,1036, 1956.

359. PARIZEK J.
The destructive effect of cadmium ion on testicular tissue and its
prevention by zinc.
U. Endocrinol. 15, 56, 1957.

360. PARIZEK J.
Sterilization of the male by cadmium salts.
J. Reprod. Fert. 1, 294, 1960.

361. PARIZEK J.

Vascular changes at sites of oestrogen biosynthesis produced by parental injection of cadmium salts.

J. Reprod. Fert. 7, 263, 1964.

362. PARIZEK J.

The peculiar toxicity of cadmium during pregnancy - an experimental toxemia of pregnancy induced by cadmium salts.

J. Reprod. Fert. 9, 111, 1965.

363. PARIZEK J.I., OSTADALOVA I., BENES I. and BABICKI A.

Pregnancy and trace elements: the protective effect of compounds of an essential trace element - selenium - against the peculiar toxic effects of cadmium during pregnancy.

J. Reprod. Fert. 16, 507, 1968.

364. PATE F.M., JOHNSON A.D. and MILLER W.J.

Testicular changes in calves following injection with cadmium chloride.

J. Anim. Sci. 31, 559, 1970.

365. PATON G.R., and ALLISON A.C.

Chromosome damage in human cell cultures induced by metal salts.

Mutat. Res. 16, 332, 1972.

366. PEDEN J.D., CROTHERS J.H., WATERFALL C.E. and BEASLEY J.

Heavy metals in Somerset marine organisms.

Mar. Pollut. Bull. 4, 7, 1973.

367. PERRY H.M. and SCHROEDER H.A.

Concentration of trace metals in urine of treated and untreated hypertensive patients compared with normal subjects.

J. Lab. Clin. Med. 46, 936, 1955.

368. PERRY H.M. and YUNICE A.

Acute pressor effects of intra-arterial cadmium and mercuric ions in anesthetized rats.

Proc. Soc. Exp. Biol. Med. 120, 805, 1965.

369. PERRY H.M. and ERLANGER M.W.

Cadmium induced increase in peripheral renin activity.

Circulation XLII, III-88, 1970.

370. PERRY H.M., ERLANGER M., YUNICE A., SCHOEPELE E. and PERRY E.F.
Hypertension and tissue metal levels following intravenous cadmium
mercury and zinc.
Am. J. Physiol. 219, 755, 1970.

371. PERRY H.M. and ERLANGER M.
Hypertension and tissue metal levels after intraperitoneal cadmium
mercury and zinc.
Am. J. Physiol. 220, 808, 1971.

372. PERRY H.M., PERRY E.F. and PURIFOY J.E.
Antinatriuretic effect of intramuscular cadmium in rats.
Proc. Soc.Exp. Biol. Med. 136, 1240, 1971.

373. PERRY H.M. and ERLANGER M.W.
Elevated circulating renin activity in rats following doses of
cadmium knowns to induce hypertension.
J. Lab. Clin. Med 82 , 399, 1973.

374. PERRY H.M. and ERLANGER M.W.
Metal induced hypertension following chronic feeding of low doses
of cadmium and mercury.
J. lab. Clin. Med. 83, 541, 1974.

375. PFEILSTICKER K. and MARKARD C.
Cadmium—Blei— und Zinkgehalt Pflanzlicher Lebensmittel aus Klein-
gärten eines Industriegebiets.
Z. Lebenzm. Unters. – Forsch. 158, 129, 1975.

376. PIOTROWSKI J.K., TROJANOWSKA B., WISNIEWSKA-KNYPL J.M. and BOLANOWSKA W.
Mercury binding in the kidney and liver of rats repeatedly exposed
to mercuric chloride; induction of metallothionein by mercury and
cadmium.
Toxicol. Appl. Pharmacol. 27, 11, 1974.

377. PIOTROWSKI J.K., TROJANOWSKA B. and SAPOTA A.
Binding of cadmium and mercury by metallothionein in the kidneys
and liver of rats following repeated administration.
Arch. Toxicol. 32, 351, 1974.

378. PISCATOR M.

Proteinuria in chronic cadmium poisoning.

I. An electrophoretic and chemical study of urinary and serum

proteins from workers with chronic cadmium poisoning.

Arch. Environ. Health 4, 607, 1962.

379. PISCATOR M.

Proteinuria in chronic cadmium poisoning.

II. The applicability of quantitative and qualitative methods of pro-

tein determination for the demonstration of cadmium proteinuria.

Arch. Environ. Health 5, 325, 1962.

380. PISCATOR M.

Proteinuria in chronic cadmium poisoning.

III. Electrophoretic and immunoelectrophoretic studies on urinary

proteins from cadmium workers with special reference to the excre-

tion of low molecular weight proteins.

Arch. Environ. Health 12, 335, 1966.

381. PISCATOR M.

Proteinuria in chronic cadmium poisoning.

IV. Gel filtration and ion exchange chromatography of urinary pro-

teins from cadmium workers.

Arch. Environ. Health 12, 345, 1966.

382. PISCATOR M. and AXELSSON B.

Serum proteins and kidney function after exposure to cadmium.

Arch. Environ. Health 21, 604, 1970.

383. PISCATOR M.

In cadmium in the environment.

Ed. Friberg, Piscator M. and Nordberg G.

CRC Press, Cleveland 1971.

384. PISCATOR M. and LIND B.

Cadmium, zinc, copper and lead in human renal cortex.

Arch. Environ. Health 24, 426, 1972.

385. PISCATOR M.

Cadmium-zinc interactions.

CEC-EPA-WHO International Symposium

Environment and Health, Paris, June 1974.

386. PISCATOR M. and LARSSON S.E.

Retention and toxicity of cadmium in calcium-deficient rats.

In Proceedings of the 17th International Congress on Occupational
Health.

(cited by Friberg et al. 129).

387. POND W.G. and WALKER E.F.Jr

Cadmium-induced anemia in growing rats: prevention by oral or
parenteral iron.

Nutr. Rep. Int. 5, 365, 1972.

388. POND W.G. and WALKER E.F.Jr.

Effect of dietary Ca and Cd level of pregnant rats on reproduction
and on dam and progeny tissue mineral concentrations.

Proc. Soc. Exp. Biol. Med. 148, 655, 1975.

389. PORTER M.C., MIYA T.S. and BOUSQUET W.F.

Cadmium: inability to induce hypertension in the rat.

Toxicol. Appl. Pharmacol. 27, 692, 1974.

390. POOTS A.M., SIMON F.P., TOBIAS J.M., POSTEL S., SWIFT M.N., PATT H.M.
and GERARD R.W.

Distribution and fate of cadmium in the animal body.

Ind. Hyg. Occup. Med. 2, 175, 1950.

391. POTTS C.L.

Cadmium proteinuria the health of battery workers exposed to cadmium oxide dust.

Ann. Occup. Hyg. 8, 55, 1965.

392. PRIBBLE H.J. and WESWIG P.H.

Effects of aqueous and dietary cadmium on rat growth and tissue
uptake.

Bull. Environ. Cont. Toxicol. 9, 271, 1973.

393. PRIGGE E., BAUMER H.P., HEERING H. and HOCHRAINER D.

Physiologische und verhaltenphysiologische Untersuchungen an Ratten
nach chronischer Inhalation von Blei- und Cadmiumhaltigen Aerosolen
allein und in Kombination mit Kohlenmonoxid unter besonderer Berück-
sichtigung der Wirkung auf den Embryo.

Report to CEC - Feb. 1976.

394. PRINCI F.

A study of industrial exposures to cadmium.

J. Ind. Hyg. Toxicol. 29, 315, 1947.

395. PRINCI F. and GEEVER E.F.

Prolonged inhalation of cadmium

Arch. Ind. Hyg. Occup. Med. 1, 651, 1950.

396. PRINGLE B.H., HISONG D.E., KATZ E.L. and MULAWKA S.T.

Trace metal accumulation by estuarine mollusks.

J. Sanitary Eng. Div., 94, 155, 1968.

397. PRODAN L.

Cadmium poisoning.

II. Experimental cadmium poisoning.

J. Ind. Hyg. Toxicol. 14, 174, 1932.

398. PRODAN L.

Cadmium poisoning.

Thesis — Harvard School of Public Health

Boston 1931.

399. PROGRAMME CIPS - Belgium

Pollution des eaux.

Inventaire (1975).

400. PUBLIC HEALTH SERVICE DRINKING WATER STANDARD, revised 1962.

U.S. Department of Health, Education and Welfare.

Public Health Service, Washington DC 1962.

401. PUJOL M., ARLET J., BOLLINELLI R. and CARLES P.

Tubulopathie des intoxications chroniques par le cadmium.

Arch. Mal. Prof. 31, 637, 1970.

402. PULIDO P., KAGI J.H.R. and VALLEE B.L.

Isolation and some properties of human metallothionein.

Biochemistry, 5, 1768, 1966.

403. RAHOLA T., AARAN R.K. and MIETTINEN J.K.

Half—times studies of mercury and cadmium by whole body counting.

In Assessment of Radioactive contamination in man.

IAEA, Vienna 1972.

404. RASTOGI R.B. and SINGHAL R.L.
     Effect of chronic cadmium treatment on rat adrenal catechlolamines.
     End. Res. Comm. 2, 87, 1975.

405. RATKOWSKY D.A., THROWEP S.J., EUSTACE I.J. and OLLEY J.
     A numerical study of the concentration of some heavy metals in
     Tasmanian oysters.
     J. Fish. Res. Board Can. 31, 1165, 1974.

406. RAUTU R. and SPORN A.
     Beitrage zur Bestimmung der Cadmiumzufuhr durch Lebensmittel.
     Nahrung 14, 25, 1970.

407. REDDI J., SVOBODA D., AZARNOFF D. and DAWAR R.
     Cadmium induced Leydig cell tumors in rat testis: morphologic and
     cytochemical study.
     J. Nat. Canc. Inst. 51, 891, 1973.

408. REYNOLDS C.V. and REYNOLDS E.B.
     Cadmium in crabs and crabmeat.
     Assoc. Public Analysts J. 9, 112, 1971.

409. ROBERTS S.A., MIYA T.W. and SCHNELL R.C.
     Tolerance development to cadmium-induced potentiation of drug
     action in male rats.
     Abstract N° 153, p. 127 in Abstracts of papers.
     Society of Toxicology. Fifteenth Annual Meeting.
     Atlanta, Georgia, 1976.

410. ROE F.J.C., DUKES C.F., CAMERON K.M., PUGH R.C.B. and MITCHLEY B.C.V.
     Cadmium neoplasia: testicular atrophy and Leydig cell hyperplasia
     and neoplasia in rats and mice following subcutaneous injection of
     cadmium salts.
     Brit. J. Cancer 18, 674, 1964.

411. ROELS H.A., LAUWERYS R.R., BUCHET J.P. and MATERNE D.
     Study on cadmium proteinuria. Glomerular dysfunction: an early
     sign of renal impairmant.
     CEC-EPA-WHO International Symposium. Paris, June 1974.

412. ROELS H.A., BUCHET J.P., LAUWERYS R.R. and SONNET J.

Comparison of in vivo effect of inorganic lead and cadmium on glutathione reductase system and $\mathcal{S}$-aminolevulinate dehydratase in human erythrocytes.

Brit. J. Industr. Med. 32, 181, 1975.

413. ROHR G. and BAUCHINGER M.

Chromosome analyses in cell cultures of the chinese hamster after application of cadmiumsulphate.

Mut. Res. 40, 125, 1976.

414. ROSMANITH J., EINBRODT H.J. and GORDON Th.

Beziehungen zwischen Blei- und Zinkniederschlägen und den Schwermetallgehalten (Pb, Zn, Cd) im Blut, Urin und in den Haaren bei Kindern.

Zbl. Bakt. Hyg. I. Abt. Orig. B. 161, 125, 1975.

415. ROSMANITH J., EINBRODT H.J. and EHM W.

Zu Interaktionen zwischen Blei, Cadmium und Zink bei Kindern aus einem Industriegebiet.

Staub-Reinhaltung der Luft. 36, 55, 1976.

416. RUGSTAD H.E. and NORSETH T.

Cadmium resistance and content of cadmium- binding protein in cultured human cells.

Nature 257, 136, 1975.

417. SABBIONI E. and MARAFANTE E.

Heavy metals in rat liver cadmium binding protein.

Environ.Physiol. Biochem. 5, 132, 1975a.

418. SABBIONI E. and MARAFANTE E.

Accumulation of cadmium in rat liver cadmium binding protein following single and repeated cadmium administration.

Environ. Physiol. Biochem. 5, 465, 1975b.

419. SANSONI B., KRACKE W., RINGER H., SCHMIDT W., DIETL F., FISCHER J. and KREUZER W.

Der Cadmiumgehalt ausgewählter Umweltproben 1971/1973.

in European Colloquium Problems of the Contamination of man and his environment by mercury and cadmium.

Luxembourg, July 1973.

420. SCHARPF L.G., HILL I.D., WRIGHT P.L., PLANK J.B., KEPLINGER M.L.
and CALANDRA J.C.
Effect of sodium nitrilotriacetate on toxicity, teratogenicity and
tissue distribution of cadmium.
Nature 239, 231, 1972.

421. SCHMIDT P., GOHLKE R., KOHLER R. and WOLFF R.
Zur akuten Inhalativen Toxizität von Kadmium stearat.
Z. Ges. Hyg. 17, 308, 1971.

422. SCHROEDER H.A. and BALASSA J.J.
Abnormal trace metals in man: cadmium
J. Chronic. Dis. 14, 236, 1961.

423. SCHROEDER H.A., VINTON W.H.Jr. and BALASSA J.J.
Effect of chromium, cadmium and other trace metals on the growth
and survival of mice.
J. Nutr. 80, 39, 1963.

424. SCHROEDER H.A., BALASSA J.J. and VINTON W.H.Jr.
Chromium, lead, cadmium, nickel and titanium in mice: effect on
mortality, tumors and tissue levels.
J. Nutr. 83, 239, 1964.

425. SCHROEDER H.A.
Cadmium hypertension in rats.
Am? J. Physiol. 207, 62, 1964.

426. SCHROEDER H.A., BALASSA J.J. and VINTON W.H.Jr.
Chromium, cadmium and lead in rats: effects on life span, tumors and
tissue levels.
J. Nutr. 86, 51, 1965.

427. SCHROEDER H.A.
Cadmium as a factor in hypertension.
J. Chronic Dis. 18, 647, 1965.

428. SCHROEDER H.A.
Municipal drinking water and cardiovascular death rates.
JAMA 195, 81, 1966.

429. SCHROEDER H.A., KROLL S.S., LITTLE J.W., LIVINGSTON P.O. and MYERS M.A.G.
Hypertension in rats from injection of cadmium.
Arch. Environ. Health 13, 788, 1966.

430. SCHROEDER H.A.
Cadmium chromium and cardiovascular disease.
Circulation 35, 570, 1967.

431. SCHROEDER H.A. and BUCKMAN J.
Cadmium hypertension.
Arch. Environ. Health 14, 693, 1967.

432. SCHROEDER H.A., NASON A.P., TIPTON I.H. and BALASSA I.J.
Essential trace metals in man: zinc - relation to environmental
cadmium.
J. Chron. Dis. 20, 179, 1967.

433. SCHROEDER H.A., NASON AP. and MITCHENER M.
Action of a chelate of zinc on trace metals in hypertensive rats.
Amer. J. Physiol. 214, 796, 1968.

434. SCHROEDER H.A., NASON A.P., PRIOR R.E., REED J.B. and HAESSLER W.T.
Influence of cadmium on renal ischemic hypertension in rats.
Amer. J. Physiol. 214, 469, 1968.

435. SCHROEDER H.A. and VINTON W.H.
Hypertension induced in rats by small doses of cadmium.
Am. J. Physiol 202, 515, 1969.

436. SCHROEDER H.A. AND NASON A.P.
Trace metals in human hair.
J. Invest. Dermatol. 53, 71, 1969.

437. SCHROEDER H.A., BAKER J.T., HANSEN N.M.Jr et al.
Vascular reactivity of rats altered by cadmium and a zinc chelate.
Arch. Environ. Health 21, 609, 1970.

438. SEMBA R., OHTA K. and YAMAMURA H.
Low-dose preadministration of cadmium levels prevents cadmium-indu-
ced exencephalia (in mice).
Teratology, 10, 96, 1974.

439. SETCHELL B.P. and WAITES G.M.H.
Changes in the permeability of the testicular capillaries and of the "blood-testis barrier" after injection of cadmium chloride in the rat.
J. Endocrinol 47, 81, 1970.

440. SHAIKK Z.A. and LUCIS O.J.
Induction of cadmium-binding protein.
Fed. Proc. 29, 298, 1970.

441. SHAIKK Z .A. and LUCIS O.J.
Biological differences in cadmium and zinc turnover.
Arch. Environ. Health 24, 410, 1972.

442. SHARRETT A.R. and FEINLEIB M.
Possible toxic water factor in coronary heart-disease.
The Lancet, July 12, 76, 1975.

443. SHIEGEMATSU I. and HASEGAWA Y.
Itai-Itai (Ouch-Ouch) disease and cadmium pollution in Japan.
United States - Japan cooperative Science Program, Conference Effects Environmental Trace Metals, Human Health, Honolulu, Hawaii, 1971.

444. SHIGEMATSU I.
Prevalence of proteinuria and glycosuria in the cadmium "polluted" (P), "slightly polluted" (SP) and "not polluted (N) areas in Japan.
Report Cadmium Research.
Committee of Japanese Public Health Association - July 1975.

445. SHIRAISHI Y. and YOSIDA T.H.
Chromosomal abnormalities in cultured leucocytes cells from Itai-Itai disease patients.
Proc. Jap. Acad. 48, 248, 1972.

446. SHIRAISHI Y., KURAHASHI H. and YOSIDA T.H.
Chromosomal aberrations in cultured human leucocytes induced by cadmium sulfide.
Proc. Jap. Acad. 48, 133, 1972.

447. SHIROISHI K., TANALKA H., ANAYAMA M. and KUBOTA K.
     Normal value of urinary cadmium.
     Jap. J. Clin. Pathol. Suppl. 19(8), 190, 1971.

448. SHUMAN M.S., VOORS A.W. and GALLAGHER P.N.
     Contribution of cigarette smoking to cadmium accumulation in man.
     Bull. Environ. Cont. Toxicol. 12, 570, 1974.

449. SILVEG W.D.
     Occurrence of selected minor elements in the waters of California.
     U.S. Geological survey water supply papger 1535-L, U.S. Government
     Printing Office, Washington D.C. 1967.

450. SINGHAL R., MERALI Z., KACEW S., SUTHERLAND D.J.B.
     Persistence of cadmium-induced metabolic changes in liver and kidney.
     Science, 183, 1094, 1974.

451. SKIG E. and WAHLBERG J.E.
     A comparative investigation of the percutaneous absorption of metal
     compounds in the guinea pig by means of the radioactive isotopes:
     ^{51}Cr, ^{58}Co, ^{65}Zn, $^{110}Ag^m$, $^{115}Cd^m$, ^{203}Hg.
     J. Invest. Dermatol. 43, 187, 1964.

452. SMITH J.C., KENCH J.E. and LANE R.E.
     Determination of cadmium in urine and observations on urinary cad-
     mium and protein excretion in men exposed to cadmium oxide dust.
     Biochem. J. 61, 698, 1955.

453. SMITH J.C. and KENCH J.E.
     Observations on urinary cadmium and protein excretion in men expo-
     sed to cadmium oxide dust and fume.
     Brit. J. Ind. Med. 14, 240, 1957.

454. SMITH J.C., WELLS A.R. and KENCH J.E.
     Observations on the urinary protein of men exposed to cadmium dust
     and fume.
     Brit. J. Ind. Med. 18, 70, 1961.

455. SMITH J.P., SMITH J.C. and McCALL A.J.
     Chronic poisoning from cadmium fume.
     J. Pathol. Bacteriol. 80, 287, 1960.

456. SNIDER G.L., HAYES J.A., KORTHY A.L. and LEWIS G.P.
Centrilobular emphysema experimentally induced by cadmium chloride aerosol.
Amer. Rev. Resp. Dis. 108, 40, 1973.

457. SPOLYAR L.W., KEPPLER J.E. and PORTHER H.G.
Cadmium poisoning in industry: report of 5 cases including one death.
J. Industr. Hyg. 26, 232, 1944.

458. SPOMER L.A.
Fluorescent particle atmospheric tracer: toxicity hazard.
Atmospheric environment, 7, 353, 1973.

459. SPORN A., DINU I. and STOENESCU L.
Influence of cadmium administration on carbohydrate and cellular energetic metabolism in the rat liver.
Rev. Roum. Biochim. 7, 299, 1970.

460. SQUIBB K.S. and COUSINS R.J.
Control of cadmium binding protein synthesis in rat liver.
Environ. Physiol. Biochem. 4, 24, 1974.

461. STATE OF CALIFORNIA. DEPARTMENT OF PUBLIC HEALTH.
Occupational health aspects of cadmium inhalation poisoning with special reference to welding and silver brazing.
September 1969.

462. STEGAVIK K.
An investigation of heavy metal contamination of drinking water in the city of Trondheim, Norway.
Bull. Environ. Cont. Toxicol. 14, 57, 1975.

463. STENSTRÖM T.
Cadmium availability to wheat: a study with radioactive tracers under field conditions.
AMBIO 3, 87, 1974.

464. STENSTROM T. and WAHTER M.
Cadmium and lead in Swedish commercial fertilizers.
AMBIO 3, 91, 1974.

465. STOKINGER H.E.

    Chapter XXVII in industrial hygiene and toxicology, vol. II.
    F.A. Patty Ed. Interscience, NY, 1963.

466. STOWE H.D., WILSON M. and GOYER R.A.

    Clinical and morphological effects of oral cadmium toxicity in
    rabbits.
    Arch. Pathol. 94, 389, 1972.

467. STOWE H.D., GOYER R.A., MEDLEY P. and CATES M.

    Influence of dietary pyridoxine on cadmium toxicity in rats.
    Arch. Environ. Health 28, 209, 1974.

468. SUDA T., HORIUCHI N., OGATA E., EZAWA I., OTAKI N. and KUMURA M.

    Prevention by metallothionein of cadmium-induced inhibition of
    vitamin D activation reaction in kidney.
    FEBS Letter 42, 23, 1974.

469. SUGAWARE N. and SUGAWARA C.

    Cadmium accumulation in organs and mortality during a continued
    oral uptake.
    Arch. Toxicol. 32, 297, 1974.

470. SUGAWARA N. and SUGAWARA C.

    Effect of cadmium, in vivo and in vitro, on intestinal brush border
    ALPase and ATPase.
    Bull. Environ. Cont. Toxicol. 14, 653, 1975.

471. SUGAWARA C. and SUGAWARE N.

    The inductive effect of cadmium on protein synthesis of rat intestine.
    Bull. Environ. Cont. Toxicol. 14, 159, 1975.

472. SUMINO K., HAYAKAWA K., SHIBATA T. and KITAMURA S.

    Heavy metals in normal Japanese tissues.
    Arch. Environ. Health 30, 487, 1975.

473. SUTER K.E.

    Studies on the dominant-lethal and fertility effects of the heavy
    metal compounds methylmercuric hydroxide, mercuric chloride, and
    cadmium chloride in male and female mice.
    Mutation Res. 30, 365, 1975.

474. SUZUKI S., SUZUKI T. and ASHIZAWA M.

Proteinuria due to inhalation of cadmium stearate dust.

Ind. Health 3, 73, 1965.

475. SUZUKI S., TAGUCHI T. and YOKOHASHI G.

Dietary factors influencing upon the retention rate of orally administered $^{115m}CdCl_2$ in mice with special reference to calcium and
protein concentration in diet.

Ind. Health 7, 155, 1969.

476. SUZUKI S. and TAGUCHI T.

Sex difference of cadmium content in spot urines.

Ind. Health 8, 150, 1970.

477. SYVERSON T.L.M.

Cadmium-binding in human liver and kidney.

Arch. Environ. Health 30, 158, 1975.

478. SZADKOWSKI D., SCHULTZE H., SCHALLER K.H. and LEHNERT G.

Zur ökologischen Bedeutung des Schwermetallgehaltes von Zigaretten.

Arch. Hyg. Bakteriol. 153, 1, 1969.

479. SZADKOWSKI D., SCHALLER K.H. and LEHNERT G.

Renale Cadmiumausscheidung, Lebensalter und Arterieller Blutdruck.

Z. Klin. Chem. Klin. Biochem. 7, 551, 1969.

480. SZADKOWSKI D.

Cadmium – eine ökologische Noxe am Arbeitsplatz.

Med. Monatsschr. 26, 553, 1972.

481. SZADKOWSKI D.

Pathogenetische une pathophysiologische Mechanismen nach Inkorpo-
ration' von Cadmium.

in European Colloquium: Problems of contamination of man and his
environment by mercury and cadmium, Luxembourg, 1973.

482. TASK GROUP ON LUNG DYNAMICS.

Deposition and retention models for internal dosimetry of the human
respiratory tract.

Health. Phys. 12, 173, 1966.

483. TASK GROUP ON METAL ACCUMULATION.
    Accumulation of toxic metals with special reference to their absorp-
    tion, excretion and biological half-times.
    Environ. Physiol. Biochem. 3, 65, 1973.

484. TAYLOR D.J.
    Report on a survey of fish products for metallic contamination under-
    taken in south west England and south Wales during early 1971.
    J. Assoc. Public. Anal. 9, 76, 1971.

485. TECULESCU D.B. and SANESCU D.C.
    Pulmonary function in workers with chronic exposure to cadmium
    oxide fumes.
    Intern. Arch. Occup. Health 26, 335, 1970.

486. TERHAAR C.J., VIS E., RONDABUSH R.L. and FASSET D.W.
    Protective effects of low doses of cadmium chloride against subse-
    quent high oral doses in the rat.
    Tox. Appl. Pharmacol. 7, 500, 1965.

487. THE-HUNG BUI, LINDSTEN J. and NORDBERG G.F.
    Chromosome analysis of lymphocytes from cadmium workers and Itai-
    Itai patients.
    Environ. Res. 9, 187, 1975.

488. THIND G.S., BIERY D.N.,BAUM S. et al.,
    Blocking of vascular receptors by cadmium: a renal angiographic
    study.
    Circulation 40 (Suppl.3), 202, 1969.

489. THIND G.S., KARREMAN G., STEPHAN K.F. and BLAKEMORE W.S.
    Vascular reactivity and mechanical properties of normal and cad-
    mium-hypertensive rabbits.
    J. Lab. Clin. Med. 76, 560, 1970.

490. THIND G.S., STEPHAN K.F., and BLAKEMORE W.S.
    Inhibition of vasopressor responses by cadmium.
    Amer. J. Physiol. 219, 577, 1970.

491. THIND G.S., BIERY D.N. and BAUM S. et al.
Use of in vivo magnification renal arteriography to study cadmium
effects on vasoactive responses.
Radiology, 9, 279, 1971.

492. THIND G.S.
Role of cadmium on human and experimental hypertension.
J. Air Pollution Control Assoc. 22, 267, 1972.

493. THIND G.S., BIERY D.N. and BOVEE K.C.
Production of arterial hypertension by cadmium in the dog.
J. Lab. Clin. Med. 81, 549, 1973.

494. THIND G.S. and FISCHER G.M.
Cadmium and zinc distribution in cardiovascular and other tissues
of normal and cadmium-treated dogs.
Exp. Mol. Pathol. 22, 326, 1975.

495. THOMAS B., ROUGHAN J.A. and WATTERS E.D.
Lead and cadmium content of some vegetable foodstuffs.
J. Sci. Food Agric. 23, 1493, 1972.

496. THOMAS B.J., HARVEY T.C., McLELLAN J.S., FREMLIN J.H. and DYKES P.W.
The measurement of liver-cadmium in patients by neutron activation
analysis. p.155 in Radioaktive Isotope in Klinik und Forschung 12.
Band.
Gasteiner Internationales Symposium 1976. Verlag H. Egermann,
Vienna 1976.

497. THURAUF J., SCHALLER K.H., ENGELHAPDT E. and GOSSLER K.
Der Kadmiumgehalt der menschlichen Placenta.
Int. Arch. Occup. Environ. Health 36, 19, 1975.

498. TIPTON I.H. and STEWART P.L.
Long term studies of elemental intake and excretion of three
adult male subjects.
Developments in Appl. Spectrosc. 8, 40, 1970.

499. TOLAN A. AND ELTON G.A.
Total diet studies with special reference to mercury and cadmium
in European Colloquium: Problems of the contamination of man and his
environment by mercury and cadmium. Luxembourg, July 1973.

500. TOWSEND R.H.

A case of acute cadmium pneumonitis: lung function test during a four year follow-up.

Brit. J. Industr. Med. 25, 68, 1968.

501. TOYASHIMA I., SEINO A. and TSUCHIYA K.

Urinary amino acids in cadmium workers in inhabitants of a cadmium-polluted area and in Itai-Itai disease patients.

In "Kankyo Hoken Report N° 24", 1973. Cited by Nordberg 1975.

502. TRUHAUT R. and BOUDENE C.

Recherche sur le sort du cadmium dans l'organisme au cours des intoxications. Intérêt en médecine du travail.

Arch. Hig. Rada, 5, 19, 1954.

503. TRUFFERT L.

Le cadmium dans les aliments.

Rapport présenté à la Commission d'étude des "Substances étrangères dans les aliments".

Ann. Fals. Fraud 219, 1950.

504. TSUCHIYA K.

Proteinuria of workers exposed to cadmium fume, the relationship to concentration in the working environment.

Arch. Environ. Health. 14, 875, 1967.

505. TSUCHIYA K.

Causation of ouch ouch disease, an introductory review.

Keio J. Med. 18, 181, 1969.

506. ULANDER A. and AXELSSON O.

Measurement of blood-cadmium levels.

Lancet i, 682, 1974.

507. UNGER M. and CLAUSEN J.

Liver cytochrome P-450 activity after intraperitoneal administration of cadmium salts in the mouse.

Environ. Physiol. Biochem. 3, 236, 1973.

508. U.S. BUREAU OF MINES

Mineral Yearbook for 1971, vol. 1, Minerals, metals, facts.

G.P.O., Washington D.C., 1973.

509. VANDER A.J.
Cadmium enhancement of proximal tubular sodium reabsorption.
Am. J. Physiol. 203, 1005, 1962.

510. VENS M.D. and LAUWERYS R.
Détermination simultanée du plomb et du cadmium dans le sang et
l'urine par le couplage des techniques de chromatographie sur
résine échangeuse d'ions et de spectrophotométrie d'absorption atomique.
Arch. Mal. Prof. 33, 97, 1972.

511. VERSAR INC.
Microeconomic analysis of the usage of various toxic substances.
EPA Contract N° 68-01-2926, Progress Report N° 1 to 5.
Springfield 1974.

512. VIGLIANI E.C., PERNIS B. and AMANTE L.
Etudes biochimiques et immunologiques sur la nature de la protéi-
nurie cadmique.
Med. Lav. 57, 321, 1966.

513. VOORS A.W., SHUMAN M.S. and GALLAGHER P.N.
Zinc and cadmium autopsy levels for cardiovascular disease in geo-
graphical context.
in Sixth Annual Conference of trace substances in environmental
Health, University of Missouri. Columbia, Mo, 1972,
(Hemphill D.D. Ed.)

514. VOROBJEVA R.S.
On occupational lung disease in prolonged action of aerosol of
cadmium oxide.
Arch. Pathologie (8, 25, 1957a) (in Russian) cited by Friberg et al.
(129).

515. VOSTAL J.J. and CHERIAN M.G.
Biliary excretion of cadmium in the rat.
Tox. Appl. Pharmacol. 29, 141, 1974.

516. VUORI E., HUUMAN-SEPPALA A. and KILPIO J.O.
The concentrations of copper, iron, manganese, zinc and cadmium in
human hair as a possible indicator of their tissue concentrations.
CEC-EPA-WHO International Symposium - Environment and Health -
Paris, June 1974.

517. WATANABE H.

A study of health indices in population in cadmium-polluted areas,
presented at meeting on Research on Cadmium Poisoning.
Tokyo, March 25, 1973 (cited by Friberg et al129).

518. WATANABE H. and MURAYAMA H.

A study on health effect indices concerning population in cadmium-
polluted area.
CEC-EPA-WHO International symposium environment and health.
Paris, June 1974.

519. WATERS M.D., CARDNER D.E., ARANYS C. and COFFIN D.L.

Metal toxicity for rabbit alveolar macrophages in vitro.
Environ. Res. 9, 32, 1975.

520. WEBB M.

Binding of cadmium ions by rat liver and kidney.
Biochem. Pharmacol. 21, 2751, 1972.

521. WEBB M.

Protection by zinc against cadmium toxicity.
Biochem. Pharmacol. 21, 2767, 1972.

522. WEBB M.

Biochemical effects of Cd^{2+} injury in the rat and mouse testis.
J. Reproduc. Fertil. 30, 83, 1972.

523. WEBB M. and DANIEL M.

Induced synthesis of metallothionein by pig kidney cells in vitro
in response to cadmium.
Chem. Biol. Interactions 10, 269, 1975.$\frac{1}{4}$

524. WEBB M. and STODDART R.W.

Isoelectric focusing of the cadmium ion-binding protein of rat liver:
interaction of the protein with a glycosaminoglycan.
Bioch. Soc. Transactions 2, 1246, 1974.

525. WEBB M. and VERSCHOYLE R.D.

An investigation of the role of metallothioneins in protection
against the acute toxicity of the cadmium ion.
Biochem. Pharmacol. 25, 673, 1976.

526. WESTCOTT D.T. and SPINCER D.

The cadmium, nickel and lead content of tobacco and cigarette smoke.
Beiträge zur Tabakforschung, 7, 217, 1974.

527. WESTER P.O.

Trace element balances in relation to variations in calcium intake.
Atherosclerosis, 20, 1974.

528. WHITE I.G.

The toxicity of heavy metals to mammalian spermatozoa.
Aust. J. Exp. Biol. Med. Sci. 33, 359, 1955.

529. WILCOX SL.

Presumed safe ambient air quality levels for selected potentially
hazardous pollutants.
Contract EPA 68-01-0438.
The Mitre Corporation, Washington, May 1973.

530. WILLDEN E.G.

Urinary and whole blood cadmium concentrations in renal disease.
Ann. Clin. Biochem. 10,107, 1973.

531. WILLIAMS C.H. and DAVID D.J.

The effect of superphsophate on the cadmium content of soils and
plants.
Aust. J. Soil. Res. 11, 43, 1973.

532. WILSON R.H., De EDS F. and COX A.J.

Effects of continued cadmium. feeding.
J. Pharmacol. Exp. Ther. 71, 222, 1941.

533. WINSTON R.M.

Cadmium fume poisoning.
Brit. Med. J., May 15, 401, 1975.

534. WISNIEWSKA-KNYPL J.M. and JABLONSKA J.

Selective binding of cadmium in vivo on metallothionein in rat
liver.
Bulletin de l'Académie polonaise des sciences
Série des sciences biologiques, 38, 321, 1970.

535. WISNIEWSKA-KNYPL J.M., JABLONSKA J. and MYSLAK Z.
Binding of cadmium on metallothionein in man: an analysis of a
fatal poisoning by cadmium iodide.
Arch. Toxicol. 28, 46, 1971.

536. WOLKOWSKI
Differential cadmium-induced embryotoxicity in two inbred mouse strains.
1. Analysis of inheritance of the response to cadmium and of the
presence of cadmium in fetal and placental tissues.
Teratology 10, 243, 1974.

537. WORKING GROUP ON HAZARDS TO HEALTH AND ECOLOGICAL EFFECTS OF METALS
AND METALLOIDS IN THE ENVIRONMENT.
Cadmium in man's environment sources, turnover and ecological effects.
Stockholm 29 October - 2 November 1973.

538. WHO
International standards for drinking water.
Third Edition, Geneva 1971.

539. WHO - REGIONAL OFFICE FOR EUROPE
The hasard to Health of persistent substances in water.
Copenhagen 1973.

540. YAMAGATA N. and SHIGEMATSU I.
Cadmium pollution in perspective.
Bull. Inst. Public Health 19, 1, 1970.

541. YAMAGATA N., IWASHIMA K., KUZUHARA Y. and YAMAGATA T.
A model surveillance for cadmium pollution.
Bull. Inst. Public health (Tokyo, 20, 170, 1971).

542. YAMAGATA N. and IWASHIMA S.
Average cadmium intake of Japanese people in Kankyo Hoken
Report N° 24
Japanese Association of Public Health, September 1973.
Cited by Friberg et al (129).

543. YEAGER D.W., CHOLAK J. and MEINERS B.G.
The determination of cadmium in biological and related material
by atomic absorption.
Amer. Ind. Hyg. Assoc. J. 450, October 1973.

544. YOSHIKAWA H. and OHSAWA M.
Protective effect of phenobarbital on cadmium toxicity in mice.
Toxicol. Appl. Pharmacol. 34, 517, 1975.

545. YOSHIKI S., YANAGISAWA T., KIMURA M., OTAKI N., SUZUKI M. and SUDA T.
Bone and kidney lesions in experimental cadmium intoxication.
Arch. Environ. Health 30, 559, 1975.

546. YUHAS E.M., MIYA T.S. and SCHNELL R.C.
Cadmium-induced alterations in calcium disposition.
Abstract N° 159, p. 131 in Abstract of papers.
Society of Toxicology.
Fifteenth Annual meeting, Atlanta Georgia 1976.

547. ZAVON M.R. and MEADOWS D.
Vascular sequelae to cadmium fume exposure.
Amer. Ind. Hyg. Assoc. J. 31, 180, 1970.

548. ZELLER H. and PEH J.
Bericht über die Prüfung von Cadmiumchlorid auf praenatale Toxizität
an Mäusen nach oraler Applikation.
BASF Medizinisch- Biologisch Forschungslaboratorium.
Gewerbehygiene und Toxikologie 1975.

549. ZIELHUIS R.L.
Biological Quality guide for inorganic lead.
Int. Arch. Arbeitsmed. 32, 103, 1974.

550. ZOOK E.G., GREENE F.E. and MORRIS E.R.
Nutrient composition of selected wheats and wheat products.
VI. Distribution of manganese, copper, nickel, zinc, magnesium,
lead, tin, cadmium, chromium and selenium as determined by atomic
absorption spectroscopy and colorimetry.
Cereal Chem. 47, 720, 1970.

551. ZOETMAN B.C.J. and BRINKMANN F.J.J.

   Human intake of minerals from drinking water in the European
   Communities.

   in European colloquium "Hardness of drinking water and public
   health" Luxembourg 21 - 23 May 1975.

ADDITIONAL REFERENCES

552. BARTHELEMY P. and MOLINE R.

   Intoxication chronique par l'hydrate de cadmium, son signe précoce:
   la bague jaune dentaire.

   Paris Médical 7, Janvier 1946.

553. BERNARD A., ROELS H.A., BUCHET J.P. and LAUWERYS R.

   α 1-antitrypsin in workers exposed to cadmium.

   Presented at Toxic metals symposium on clinical chemistry and chemical toxicology.

   Monte-Carlo, Monaco 2-5 March, 1977.

554. BOUDENE C.

   Recherches toxicologiques sur le cadmium.

   Etude analytique et biologique. Thèse. Université de Paris, 1955.

555. DORN C.R., PIERCE J.O., PHILLIPS P.E. and CHASES G.R.

   Airborne Pb, Cd, Zn and Cu concentration by particle size near a
   Pb smelter.

   Atmospheric Environment 10, 443, 1976.

556. FABRICIUS G., WALBER K. and HILSCHER W.

   Störungen der Embryonalenentwicklung durch Cadmium.

   Paper presented at the 6th International Hygienetagung,
   Mainz, 1976.

557. GOYER R.A., CHERCARI G.M. and RICHARDSON L.D.

   The pathogenesis of cadmium nephropathly

   Paper presented at the First international cadmium conference
   San Francisco, 31 January - 2 February 1977.

558. ISHIZU S., MIANAMI M., SUZUKI A., YAMADA M., SATO M. and YAMAMURA K.
An experimental study on teratogenic effects of cadmium.
Indust. Health 11, 127, 1973.

559. KJELLSTRÖM T.,
Accumulation and renal effects of cadmium in man. A dose-response study.
Doctoral thesis
Department of Environ. Hyg. Karolinska Institute, Stockholm, 1977.

560. VON RÖPENACK
Practical emission levels in the production of cadmium metal.
Paper presented at the First International cadmium conference
San FRancisco, 31 January - 2 February 1977.

ACKNOWLEDGEMENTS

The Commission of the European Communities wishes to acknowledge the help given by the scientific experts and consultants who advised the Health and Safety Directorate in drawing up this report.

COUNTRY	NAME	ADDRESS	MEETING NUMBER:
BELGIUM	P. BRUAUX	Institut d'Hygiène et d'Epidémiologie 14 rue Juliette Wytsman 1050 Bruxelles	1, 2
	R. LAUWERYS	Université Catholique de Louvain 30 avenue Chapelle-aux-Champs 1200 Bruxelles	1, 2
	L. PATERNOTTE	Ministère de l'Emploi et du Travail 53 rue Belliard 1040 Bruxelles	1, 2
	R. F. RECKERS	Ministère de la Santé Publique Cité Administrative Quartier Vésale Bruxelles	1
DENMARK	J. CARSTENSEN	Institute of Toxicology National Food Institute 19 Mørkhøj Bygade 2860 Søborg	2
	N. CHRISTENSEN	Danish Agency for the Environment 1 Kampmannsgade Copenhagen V	2
	A. GRUT	State Labour Inspection Hambros Alle 20 2900 Hellerup	1, 2
	J. C. HANSEN	Institute of Hygiene University of Aarhus 8000 Aarhus	1, 2
	E. POULSEN	Institute of Toxicology National Food Institute 19 Mørkøj Bygade 2860 Søborg	1

F.R. GERMANY	G. CROSSMANN	Landwirtschaftliche Untersuchungs- und Forschungsanstalt Von Essmarchstrasse 12 44 Munster	1
	A. KLOKE	Biologische Bundesanstalt Königin-Luise Strasse 19 1000 Berlin 33	1, 2
	D. LORKE	Institut für Toxikologie der Bayer AG Friedrich-Ebert Strasse 217 56 Wuppertal 1	1, 2
	C. MARKARD	Umweltbundesamt Bismarkplatz 1 1 BERLIN 33	1, 2
	H. OLDIGES	Institut für Aerobiologie der Fraunhofer-Gesellschaft 5948 Schmallenberg-Grafschaft	1, 2
	D. RADE	Bayer AG Rheinuferstrasse 7-9 4150 Krefeld 11	1
	H.-W. SCHLIPKOETER	Medizinisches Institut für Lufthygiene Universität Düsseldorf Gurlittstrasse 53 4 Düsseldorf	1, 2
	E. SCHMIDT	Bundesgesundheitsamt - ZEBS Postfach 1 Berlin 33	1, 2
	N. SEEMAYER	Medizinisches Institut für Lufthygiene Universität Düsseldorf Gurlittstrasse 53 4 Düsseldorf	1
FRANCE	C. BOUDENE	INSERM - Laboratoire de Toxicologie Faculté de Pharmacie Paris-Sud rue J.-B. Clément 92290 Chatenay-Malabry	1, 2
	B. FESTY	Lab. d'Hygiène de la Ville de Paris 1 bis rue des Hospitalières St. Gervais 75004 Paris	1
	M. JOUAN	Ministère de la Santé Direction Générale de la Santé 20 due d'Estrée 75700 Paris	1, 2

FRANCE (continued)	P. LAUGEL	Faculté de Pharmacie Université Louis Pasteur 3 Rue de l'Argonne Strasbourg	1
	J. QUERSONNIER	D.I.C.T.D. Ministère de l'Industrie 66 rue de Bellechasse 75700 Paris	2
ITALY	S. CERQUIGLINI MONTERIOLO	Istituto Superiore di Sanità Viale Regina Elena 299 00161 Roma	1
	F. COTTA-RAMUSINO	Istituto Superiore di Sanità Viale Regina Elena 299 00161 Roma	1
	S. FUSELLI	Istituto Superiore di Sanità Viale Regina Elena 299 00161 Roma	2
	N. SARTI	Ministero della Sanità Via dell'Elettronica 18 00144 Roma	1
NETHERLANDS	E. I. KRAJNC	Rijksinstituut voor de Volksgezondheid P.O. Box 1 2660 Bilthoven	2
	A. G. RAUWS	Rijksinstituut voor de Volksgezondheid P.O. Box 1 2660 Bilthoven	1
	E. VAN JULSINGHA	Ministry of Public Health and Environment Dr. Reyersstraat 12 Leidschendam	1, 2
UNITED KINGDOM	J. FACER	Department of Health and Social Security Hannibal House Elephant and Castle London E.C.1.	2
	J. P. GILTROW	Central Unit on Environmental Pollution Department of the Environment 2 Marsham Street London S.W.1.	2
	D. G. LINDSAY	Ministry of Agriculture, Fisheries & Food Great Westminster House Horseferry Road London S.W.1.	2

INDEPENDENT EXPERT	J.F.A. THOMAS	c/o O.E.C.D. 37 bis Boulevard Suchet 75016 Paris, France	2

UNICE	J. SPAAS	6 rue de Loxum 1000 Bruxelles Belgium	2

WORLD HEALTH ORGANIZATION	P. MACUCH	Avenue Appia 1211 Geneva 27 Switzerland	1
	J. PARIZEK	Control of Environmental Pollution Unit 1211 Geneva 27 Switzerland	2

COMMISSION OF THE EUROPEAN COMMUNITIES		Health and Safety Directorate Avenue Alcide de Gasperi Luxembourg-Kirchberg	
	A. BERLIN		1, 2
	W. J. HUNTER		1, 2
	M. LANGEVIN		2
	J. SMEETS		2
	G. TREU		1, 2
	M.Th. VAN DER VENNE		1, 2
		Environment & Consumer Protection Service 200 rue de la Loi 1049 Bruxelles, Belgium	
	M. BIART		1
	G. DEL BINO		1
	L. V. WAMBEEKE		2
		D.G. Research, Science and Education 200 rue de la Loi 1049 Bruxelles, Belgium	
	E. DI FERRANTE		1, 2

1. First Meeting of National Experts, Luxembourg, 28 & 29 January 1976

2. Second Meeting of National Experts, Luxembourg, 14 & 15 October 1976